新訂版 **New Edition**

陳粉玉 著

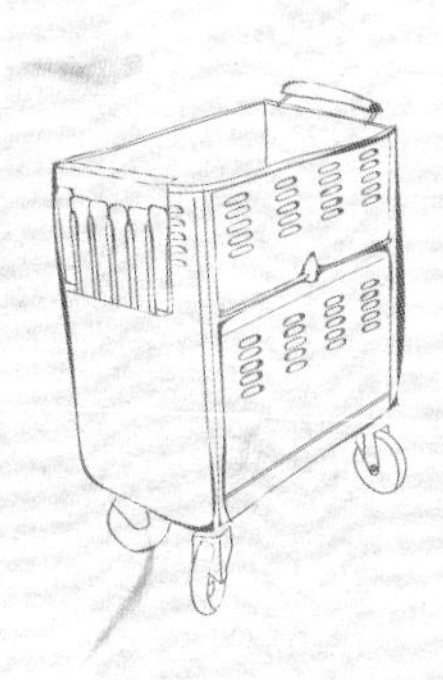

前言 Preface

「小吃」又稱小食，是正餐以外的食品，與一般佐膳的菜式比較，有份量較少、取材廣泛、食用方便和經濟實惠的特色。

見諸文字記載的小吃品種，其實可以上溯到三千年前，不過「小吃」這個詞，還是在宋朝時才出現，當時有一本名為《能改齋漫錄》的書，裏面便有「世俗例，以早晨小吃為點心」之句。發展至今，小吃已經不止作為果腹的食料，更能提供藝術欣賞和美的享受哩！

小吃的產生和流行，有一定的文化背景和歷史條件，所以中國各地的小食都各具特色，例如北方小食以麵食為主，南方則多以米為基本原料，而具有嶺南風味的廣東小吃，卻以花色繁多，做工精細而馳名。特別在昔日有「食在廣州」之稱的羊城，小食種類尤其豐富多彩，就以豬腸粉為例，原是六十年前廣州流動攤販上街叫賣的大眾小吃，後來則發展到茶樓酒家都有供應，品種也從齋腸到蝦腸、牛肉腸、叉燒腸等，配料和調料也極其講究。

香港的飲食師承自廣州，但「青出於藍而勝於藍」，如今更享有「食在香港」的美譽。香港中西文化交匯，在小吃製作上揉合了華洋特色的用料和製作手法，吃來更令人感到可喜。

香港的特色小吃可分做兩大類：一類是民間節令小吃，這些小吃通常是在家裏製作，用以應節的。因為以前的婦女大多不用上班，所以就有較多時間來做些小吃給家人、鄰居和親友品嘗。如：蘿蔔糕、八寶糯米飯、豬腳薑蛋醋、茶泡等，都是主婦們於節令期間製作的小吃。另一類是街頭小吃，這類小吃通常都是由小販挑着擔子，又或推着車子在街上聚集販賣，好像雞蛋仔、眉豆茶果、白糖糕、砵仔糕、豬皮蘿蔔、碗仔翅等。

隨着時代的演變，現今許多街頭小食都搖身一變成為茶樓酒家中的奉客佳品了。

不過，不少人都感到，在市面上吃到的許多小食不是人工色素過多，就是味精太重，吃後總覺得口裏乾巴巴的，用料和製作過程也沒有甚麼標準，要吃到好味和合乎衛生的小吃，似乎越來越難了。有見及此，我編寫了這八十八個美味而健康的小吃食譜以饗同好，好讓各位不用「偷師」也能把這些精采的小吃搬到家裏去。

有人說：「美食是很厲害的武器。」但我更認為，「美食之中的小吃是更厲害的武器。」因為無論大小朋友，都會吃得津津有味。說到製作方法，小吃也很「平易近人」，與烹製一般家庭菜式相比，小吃不需應用太多的材料配搭及太複雜的製作技巧，一般只要多試做幾次，便不難成為這方面的「師傅」。況且通過親手製作小食，不但可以作為初入廚者學習廚藝的第一步，更可促成一家大小共享天倫之樂哩！既然如此，不妨就利用餘暇，坐言起行，炮製出揉合你的心意的香港特色小吃來！

陳粉玉

（前言摘自 1998 年《香港特色小吃》初版）

Chinese snacks offer to people, in addition to the staple meal of the day, a very large variety of mini-items of food made available readily and at only commonplace costs. The term made its first appearance in publications of the Sung Dynasty (960-1279A.D.) but the history of the Chinese snacks in fact dates back to well over 3000 years ago. Today, these food specialties are not only the pursuit of city gourmets but are also regarded as a contemporary cooking art.

Chinese snacks of different geographical regions tend to reflect their cultural and historical characteristics. Snacks of North China are mainly noodles and pasta, whereas rice-derived items prevail in South China. Guangdong Province, which lies south of Nanling Range, is well known for the rich variety and the delicacy of its snacks, hence went the saying "Ideal Eating Out in Guangzhou" (Guangzhou being the Provincial capital of Guangdong). The Steamed Rice Sheet Roll is a typical southern snack. Now offered in restaurants with the vegetarian, beef, shrimp and roast pork as its popular varieties, it actually originated from hawker stalls in the streets of Guangzhou about sixty years ago.

Although Hong Kong inherits her cuisine from Guangzhou, the meeting of the Eastern and Western cultures here has led to much improvement on the preparation and serving of the traditional snacks. It is why the city has already succeeded the title of Guangzhou to achieve "Ideal Eating Out in Hong Kong".

Generally speaking, there are two kinds of snacks. The festive foods are made and served at home by housewives for season's celebrities with family members, relatives and friends. Examples are the Turnip Puddings, Eight Treasures Rice Pudding, Chinese Assorted Pickles and the Assorted New Year Crispies.

The other kind is the hawker snacks. It carries a street culture. Cooked foods are tendered by the mobile hawkers include the Chinese Egg Puff, Eye-bean Dumplings, White Sugar Sponge, Clay-pot Puddings, Soy Pig's Skin and Turnips, "Shark's Fin" Soup made from mung bean thread and many others. It is interesting to note that many of these snacks have now already got a firm place in restaurant menus.

Chinese snacks have by all means been a part of our daily lives. However, modern people are more increasingly aware of the overuse of colourants and other food additives in restaurant foods. Nor are the ingredients and preparation procedure anyhow standardized. I have endeavoured to recommend these 88 recipes to all that intend to make and enjoy their own health-and-environmental-friendly Chinese snacks at home.

Indeed, if regular dishes of Chinese cuisine are fascinating, then, its snacks are even more so. The latter usually involve simpler kitchen work and ingredients, appeal to people of different age groups. Try out Chinese Snacks may well be your first step towards professional cooking. Before long, you may become a cookery expert!

Becky Chan

(The preface was taken from *Distinctive Snacks of Hong Kong* of 1998.)

目錄
Contents

懷舊小食
Selective Oldies

「記得當時年紀小……」哼着歌兒，想起媽媽用過的木製蘿蔔絲刨；兒時，天還未亮，聽着哥哥替媽媽弄炸漿時的拍打聲；放學回家，嘗着媽媽在街頭擺賣的炸番薯、芋頭及蘿蔔絲……那一張「油淋淋」的油紙上沾着的甜醬滋味。昔日的美食與情懷，互相緊扣。還有砵仔糕、眉豆糕、西米餅、碗仔翅……此間一一呈現。

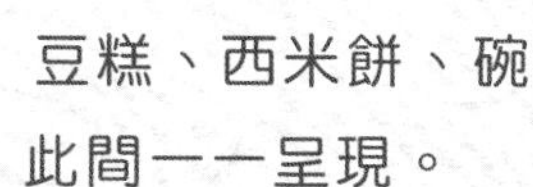

蘿蔔絲餅
Turnip Cakes

蘿蔔絲餅
Turnip Cakes

份量：8-10件 / Makes 8-10 pieces

材料

白蘿蔔 275 克
蝦米 2 湯匙
葱粒 2 棵份量
麵粉 100 克
粘米粉 40 克
發粉 1/2 茶匙
清水 250 毫升

調味料

鹽 3/4 茶匙
五香粉 1/2 茶匙
胡椒粉少許
麻油 1 茶匙

做法

1. 麵粉、粘米粉及發粉篩好。
2. 蘿蔔刨絲，蝦米浸軟，切碎，放入調味料及葱粒拌勻，置筲箕內醃片刻，使去掉多餘水分。
3. 粉料與清水調勻成粉漿，加入油、鹽及胡椒粉各少許，拌勻。
4. 燒熱一鑊油，放進長柄模型燒熱。圖 1
5. 將蘿蔔料加入粉漿內拌勻，當油燒至八成熱時，取出熱餅模，滴乾油分，加入適量蘿蔔漿，放回油中炸至離模及呈餅狀。圖 2~5
6. 將蘿蔔餅炸至鬆脆及呈金黃色，吸乾油分熱食。

心得

- 每次加入粉漿時必須將模浸入油內燒熱，蘿蔔餅才易鬆脫。

Ingredients

275 g turnip
2 tbsp dried shrimps
2 stalks spring onion
100 g plain flour
40 g rice flour
1/2 tsp baking powder
250 ml water

Seasonings

3/4 tsp salt
1/2 tsp five-spice powder
pinch of pepper
1 tsp sesame oil

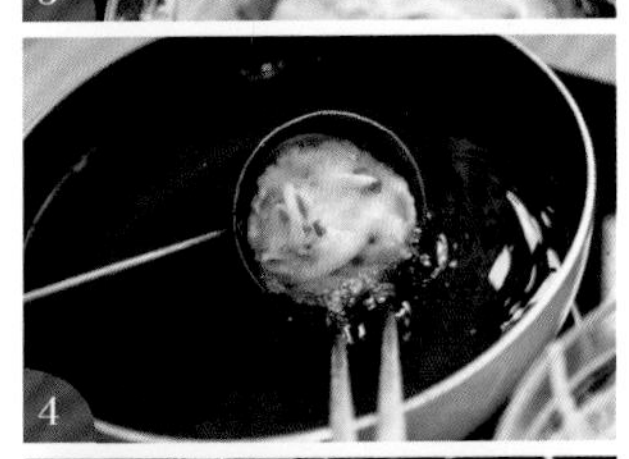

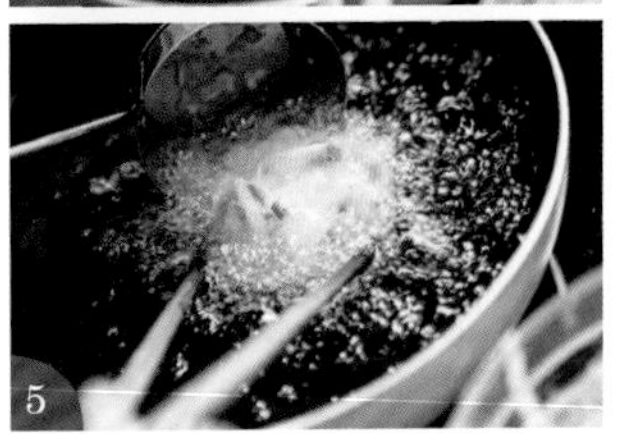

Method

1. Sieve plain flour, rice flour and baking powder together.
2. Grate the turnips; soak and dice the dried shrimps. Mix them with seasonings and diced spring onion, leave in colander to drip dry mixture.
3. Mix powdery ingredients with water to form a smooth batter, sprinkle in a little pepper, salt and oil, stir well.
4. Heat oil, slide in the metal mould until the mould is heated through. Picture 1
5. Mix the turnip mixture into the batter, lift and drain the heated mould, half-fill the mould with mixture, leave in medium-hot oil, unmould when an outer coat is formed. Pictures 2-5
6. Deep-fry turnip cakes until crispy and golden brown. Drain and serve.

Practical Tips

- The mould should be well heated in the oil between each addition of batter mixture to prevent sticking.

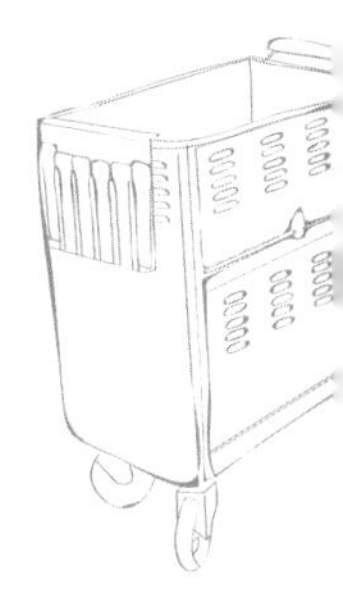

葱油餅
Spring Onion Cakes

份量：4 件 / Makes 4 pieces

材料

麵粉 150 克
暖水 125 毫升
鹽 1 茶匙
油 2 湯匙
幼葱粒 4 棵份量

Ingredients

150 g plain flour
125 ml warm water
1 tsp salt
2 tbsp oil
4 stalks spring onion, diced

做法

1. 麵粉篩至幼滑，加入暖水，攪拌，搓成幼滑粉糰，分成 4 份。
2. 枱面灑粉，用木棍將每份粉糰壓成薄長方塊。
3. 在麵皮上掃油，灑上適量幼鹽及葱粒。圖 1
4. 將兩邊同時捲向中央，覆摺成長條形，略按緊使葱粒不漏出。圖 2~5
5. 將葱油餅開口處向外捲成螺旋形，再用麵棍略壓扁。圖 6~10
6. 置中火油鍋內炸至金黃及鬆脆即成。

心得

- 麵皮壓得越薄，向內捲的層次越多，葱油餅越鬆脆。

Method

1. Sift plain flour, mix in sufficient warm water, knead to a soft dough. Divide into 4 portions.
2. On a floured pastry board, roll each piece of dough into a paper-thin rectangle.
3. Brush pastry with oil, sprinkle on salt and diced spring onion. Picture 1
4. Roll from two sides to the centre. Fold to form a long strip. Press to seal the edge. Pictures 2~5
5. Coil from folded edge to form a round cake. Flatten spring onion cakes slightly. Pictures 6~10
6. Deep-fry over medium heat until crispy and golden brown. Serve hot.

Practical Tips

- The thinner the pastry, the more the layers can be made, and the fried spring onion cakes will be more crispy.

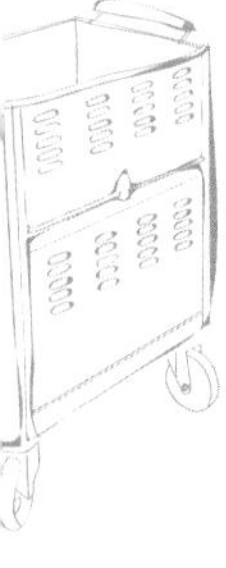

上湯煎粉粿
Crispy Dumplings (Fan Kor) in Soup

上湯煎粉粿
Crispy Dumplings (Fan Kor) in Soup

份量：6 件 / Makes 6 pieces

餡料

蝦肉 100 克
瘦肉 50 克
冬菇 2 朵
冬筍 40 克
芫茜 1 棵

粉粿皮料

澄麵粉 80 克
粟粉 1 湯匙
鹽少許
滾水 150 毫升
糖 1 茶匙
食用臭粉 1/8 茶匙
豬油 1.5 茶匙

調味料（肉粒）

鹽、糖、粟粉、生抽、麻油各少許

粟粉水

水 1 湯匙
粟粉 1 茶匙

湯料

上湯 250 毫升
韭黃 20 克

做法

1. 蝦肉切粒，瘦肉切粒，加調味料拌勻；冬菇浸軟去蒂切粒，冬筍切粒、芫茜切碎。燒油炒熟瘦肉粒，加入蝦粒炒一會，加入冬菇粒及冬筍粒，埋粟粉水，盛起，拌入芫茜碎，待涼。
2. 澄麵粉、粟粉及鹽同置深碗內，沖入大滾水，以竹筷子急速拌勻至熟，趁熱加入臭粉、豬油和糖，拌勻；搓至柔軟，分成 6 份，每份搓圓按扁，包入餡料，對摺埋口，捏成半圓形粉粿狀。
3. 用中大火油，將粉粿放入炸至浮起及呈微金黃色，撈起瀝乾油分。
4. 韭黃切粒置湯碗內，注入滾上湯，與炸好粉粿同上，熱食。

心得

- 將粉粿浸入上湯內，浸片刻才進食，取其鬆脆、香、軟、滑之口感！

Filling

100 g shelled shrimps
50 g lean pork
2 Chinese mushrooms
40 g bamboo shoot
1 stalk coriander

Dough

80 g ungluten flour
1 tbsp cornstarch
pinch of salt
150 ml boiling water
1 tsp sugar
1/8 tsp ammonia (edible)
1.5 tsp lard

Seasoning (Pork)

each of a little salt, sugar, cornstarch, light soy sauce and sesame oil

Thickening

1 tbsp water
1 tsp cornstarch

To Serve

250 ml stock
20 g yellow chives

Method

1. Dice shelled shrimps and pork, add seasonings and mix well; soak Chinese mushrooms, remove stalk and dice; dice bamboo shoot; chop coriander. Heat oil, stir-fry pork until cooked, add shrimps, stir-fry for a while, add Chinese mushrooms and bamboo shoot, thicken, dish and mix in chopped coriander. Cool.
2. Put ungluten flour, cornstarch and salt in a deep bowl, pour in boiling water, stir quickly to bind to a transparent dough, add in ammonia, lard and sugar, mix well, knead into a long roll and divide into 6 portions. Flatten each portion, wrap in filling, fold, seal edges to form a half-moon shape (Fan Kor).
3. Deep-fry Fan Kor over a medium high heat, keep turning until floating and a slightly brown. Dish and drain off excess oil.
4. Dice yellow chives and put in serving bowl, pour in boiling stock. Serve with fried Fan Kor.

Practical Tips

- To serve, dip Fan Kor in chive stock and soak for a while to get a short, crispy, fragrant, smooth and yummy feel!

生煎牛肉包
Fried Beef Buns

份量：4-5 個 / Makes 4-5 buns

餡料

絞碎牛肉 100 克
冬菜 20 克
椰菜絲（灼熟）少許

皮料

低筋麵粉 100 克
中筋麵粉 20 克
泡打粉 3 克
快速乾酵母 2 克
砂糖 10 克
油 1 茶匙
暖水 75 毫升

調味料

砂糖 1/2 茶匙
胡椒粉適量
鹽少許
老抽 1 湯匙
麻油 1 茶匙
水 2 湯匙
低筋麵粉 2 湯匙（後下）

做法

1. 牛肉加調味料拌勻，加入冬菜及椰菜絲，拌勻；最後下低筋麵粉，用手將餡料搓按片刻，置雪櫃內候用。圖 1
2. 低筋、中筋麵粉與泡打粉一同篩入大碗中，加入快速乾酵母、砂糖及油，拌勻；加入暖水，拌勻成糰，用手搓勻，置枱面或木板上搓壓至紡滑不黏手為止（約 5-8 分鐘），將麵糰放牛油紙上，置和暖蒸氣上，蓋好（發酵 30 分鐘）。
3. 取出麵糰，用拳頭壓走氣體，再壓摺成幼滑麵糰，再分成 4-5 份小粉糰，包入牛肉餡，埋口，搓圓按平，再發酵 20 分鐘。
4. 平底鑊加油少許，放入牛肉包，開火，注入清水浸至半滿（牛肉包的一半）圖 2，加蓋收小火蒸煮 15 分鐘左右，待水分收乾後開蓋，將牛肉包煎至兩面呈金黃色，熱食。

心得

- 如果使用普通的乾酵母，可先用暖水（攝氏 25-28 度）及少許砂糖浸片刻至起泡，才混合粉料中。可參考 P.219 的圖片

Filling

100 g minced beef

20 g preserved Tianjian white cabbage

a little shredded cabbage, blanch

Dough

100 g low gluten flour

20 g plain flour

3 g baking powder

2 g active dried yeast

10 g castor sugar

1 tsp oil

75 ml warm water

Seasonings

1/2 tsp sugar

shakes of pepper

pinch of salt

1 tbsp dark soy sauce

1 tsp sesame oil

2 tbsp water

2 tbsp low gluten flour, add later

Method

1. Add seasonings into minced beef, mix well, add preserved Tianjian white cabbage and blanched cabbage, mix well, sprinkle in low gluten flour, bind and press mixture with hands until well mixed, chill. Picture 1
2. Sieve low gluten flour, plain flour and baking powder together in mixing bowl, add active dried yeast, sugar and oil, mix well, add warm water to form a dough, knead thoroughly for about 5-8 minutes on floured board until smooth and not sticky. Put dough on greased proof paper, covered in a warm pan to proof for 30 minutes.
3. Take out dough, press to release the gas from the dough, fold, press and knead slightly to form a very smooth non-sticky dough, divide into 4-5 portions, wrap in beef fillings, seal and roll to a ball shape, press to flatten a bit. Proof for 20 minutes.
4. Heat pan with a little oil, add in beef buns, pour in water to half-filled the buns Picture 2, cover and cook the buns for 15 minutes over low heat. When almost dried up, remove lid and shallow fry the cooked buns until golden brown on both sides, serve hot.

Practical Tips

- If ordinary dried yeast is used (not active), add into the warm water (25-28℃) with a little sugar, stand aside until foamy before adding into the floury mixture.

眉豆茶粿
Eye Bean (Mei Dau) Dumplings

眉豆茶粿
Eye Bean (Mei Dau) Dumplings

份量：10 件 / Makes 10 pieces

材料

竹葉 5 塊
眉豆 60 克
糯米粉 150 克
暖水 150 毫升

調味料

鹽 1/2 茶匙
胡椒粉適量
糖少許
麻油 1 茶匙

做法

1. 竹葉洗淨，用熱水浸軟，剪成 10 小塊竹葉片，鋪平（平滑面向上），掃油待用。圖 1
2. 眉豆浸 1 小時左右，瀝乾，注入滾水至剛浸滿眉豆，中火燉 25-30 分鐘，倒入鑊中，加油少許，用中火炒壓至成眉豆泥，可保留部份原粒眉豆增加口感圖 2，加入調味料，拌匀，上碟，待涼。
3. 糯米粉放深碗內，加鹽、糖各少許，用暖水開成軟滑不黏手的粉糰，分成 10 份，每份搓圓按扁，包入 1 湯匙眉豆餡，包好埋口，做成扁平茶粿，放在已塗油的竹葉上，用中火蒸 10-15 分鐘，取出，待冷片刻，即可進食。

1

2

心得

- 蒸眉豆茶粿時，須中途掀起鑊蓋疏氣兩次，以免糯米糰過分受熱膨脹而變形。

Ingredients

5 pieces of bamboo leaves
60 g eye beans
150 g glutinous rice flour
150 ml warm water

Seasonings

1/2 tsp salt
shakes of pepper
pinch of sugar
1 tsp sesame oil

Method

1. Rinse bamboo leaves, soak in boiling water until quite soft, trim into 10 small bamboo sheets, lay flat (smooth sides facing upward), grease for later use. Picture 1
2. Soak eye beans for about 1 hour, drain and pour adequate amount of boiling water to just soak through the beans , steam for 25-30 minutes until cooked, stir fry in wok with a little oil, mash to form eye bean puree, add seasoning, mix well, dish and cool. Picture 2
3. Put glutinous rice flour in a deep bowl, add a pinch of salt and sugar, mix in warm water, stir and mix well to form a smooth non-sticky dough, divide into 10 portions, wrap in 1 tbsp eye-bean fillings, seal, roll and flatten, put on greased bamboo sheets, steam over a medium heat for 10-15 minutes until cooked, dish, cool down a bit before serving.

Practical Tips

- In between steaming, lift the lid twice to release excess steam from expanding in order to keep the dumplings in shapes.

豬腸粉
Steamed Rice Sheet Rolls

份量：8 條 / Makes 8 rolls

材料

粘米 225 克
清水 750 毫升
粟粉 25 克
鹽 1 茶匙
熟油 1 湯匙
葱粒 2 湯匙
蝦米 3 湯匙（浸軟、粗剁）

豉油汁

熟油 2 湯匙
上湯 2 湯匙
生抽 2 湯匙
老抽 1/2 湯匙
糖 1 茶匙

做法

1. 粘米洗淨浸 5-6 小時，加適量水分置攪拌機內磨成米漿，隔去粗粒，加入粟粉、鹽及熟油，與剩餘之水分拌勻。
2. 腸粉鍋內燒熱大半鍋水，架上鋪上一塊濕薄布，傾入米漿 1 湯杓，灑下葱粒及蝦米碎。
3. 加蓋蒸 3-5 分鐘，取出，反轉粉皮於已塗油的平面上，拉起薄布，捲成腸粉，切件上碟。
4. 煮滾豉油汁料，淋上腸粉面，趁熱享用。

心得

- 腸粉鍋可到專門售點心用具之店舖購買，或可用淺平方焗盤代替鐵架。腸粉布必須濕透及多用數次，粉皮才易脫離。

Ingredients

225 g long-grain rice
750 ml water
25 g cornstarch
1 tsp salt
1 tbsp cooked oil
2 tbsp diced spring onion
3 tbsp dried shrimps, soaked and diced

Soy Sauce

2 tbsp cooked oil
2 tbsp stock
2 tbsp light soy sauce
1/2 tbsp dark soy sauce
1 tsp sugar

Method

1. Soak the long-grain rice for 5-6 hours, add sufficient water and blend to a fine rice solution. Drain well. Add cornstarch, salt and oil, mix with the remaining water.
2. Prepare steamer, line the rack with wet muslin, spoon in a ladle of rice solution, sprinkle on diced spring onion and diced dried shrimps.
3. Cover and steam for 3-5 minutes until set. Take out and turn the rice sheet upside down on a greased smooth surface. Remove the muslin, roll to the shape of a cylinder; cut into sections.
4. Bring the ingredients of soy sauce to boil and pour onto the rice sheet rolls. Serve hot.

Practical Tips

- Special steamers for rice sheet rolls can be bought from shops that sell dim sum equipment. Or a shallow baking tin can be used. Muslin for rice rolls should be well dampened and reused a few times for better result.

雞絲粉卷
Chicken Rice Sheet Rolls

份量：6 條 / Makes 6 rolls

材料

河粉皮 1 塊
絞碎雞肉 150 克
冬菇 2 朵
薑絲 1 湯匙
芹菜絲少許

調味料

鹽、糖、胡椒粉、麻油、生粉各少許

醬汁

生抽 2 茶匙
老抽 2 茶匙
水 2 湯匙
熟油 1/2 湯匙
糖少許

做法

1. 冬菇浸軟去蒂切絲，與雞肉及薑絲拌勻，加入調味料，拌勻後放雪櫃內冷藏 30 分鐘；取出，加入芹菜絲拌勻，分成 6 份。
2. 河粉皮攤開，裁成 6 小塊長形粉皮（8x12 厘米），包入雞絲餡料，捲好，置已塗油的平碟上，中火蒸 10-15 分鐘，取出。
3. 醬汁料加熱拌勻，淋在粉卷上，熱食。

心得

- 新鮮河粉皮要預先向相熟的粉麵店訂購，或用現成齋腸粉，打開成塊狀來使用。

Ingredients

1 piece of rice sheet
150 g chicken meat, minced
2 Chinese mushrooms
1 tbsp shredded ginger
a little shredded Chinese celery

Seasonings

adequate amount of salt, sugar, pepper, sesame oil and cornstarch

Sauce Mix

2 tsp light soy sauce
2 tsp dark soy sauce
2 tbsp water
1/2 tbsp cooked oil
pinch of sugar

Method

1. Soak and remove stalk of Chinese mushrooms, shred and mix well with minced chicken, shredded ginger and seasonings. Chill for 30 minutes, take out and mix in shredded Chinese celery. Divide mixture into 6 portions.
2. Unfold rice sheet, trim into 6 pieces of rice sheets each 8x12 cm in size, add chicken fillings, roll up and fix in size. Place on greased plate and steam over medium heat for 10-15 minutes until cooked.
3. Serve hot with heated sauce mix.

Practical Tips

- Fresh rice sheet has to be pre- ordered from noodles shops, or use plain rice sheet roll (Cheung Fun), unroll and trim for wrapping.

臘腸卷
Chinese Sausage Buns

份量：4-6 個 / Makes 4-6 buns

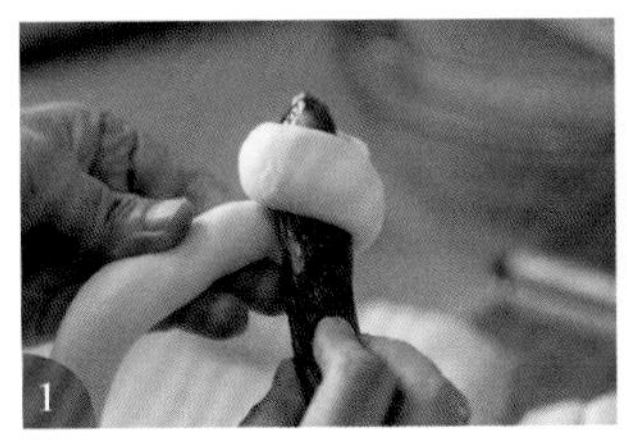
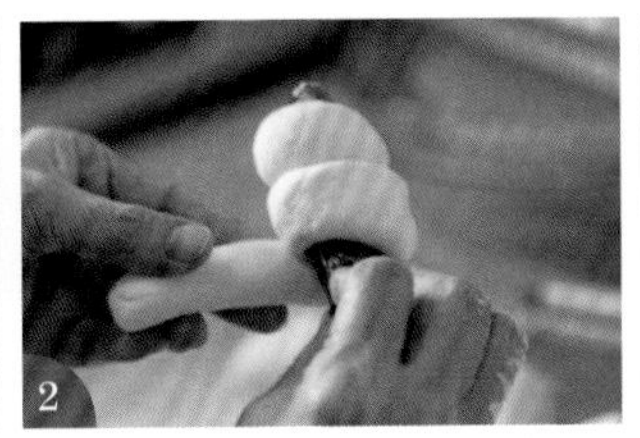

餡料

臘腸 2 條
海鮮醬 1 湯匙

皮料

低筋麵粉 120 克
澄麵粉 30 克
糖 10 克
泡打粉 3 克
快速乾酵母 2 克
油 1 茶匙
暖水 100 毫升

做法

1. 臘腸用熱水洗淨，蒸熟切成 2-3 小段，塗上海鮮醬一層，候用。
2. 低筋麵粉、澄麵粉、泡打粉與糖一起篩勻，加油及快速乾酵母，拌勻後徐徐加入暖水，拌勻成粉糰。
3. 在案板上將粉糰搓撻（約 5-8 分鐘）至幼滑及不黏手為止。
4. 將麵糰放牛油紙上，置和暖蒸氣上，蓋好，發酵 30 分鐘。
5. 取出，用拳頭壓走氣體，輕輕搓搯至軟滑，分成 4-6 份；每份搓成長條狀，繞捲入臘腸，末段收口，放在餅底紙上，靜置一旁，再發酵 15 分鐘。圖 1-3
6. 大火蒸 10 分鐘，即成。

心得

- 乾酵母可用暖水及少許糖混合，當起泡後才倒入粉料中。若使用快速乾酵母 (active yeast) 則可直接與粉料混合使用。
- 麵糰第一次發酵時要觀察其發麵速度。蒸氣過熱或發酵過久都會影響發酵效果。

Filling

2 pieces Chinese sausage
1 tbsp Hoi Sin sauce

Dough

120 g low gluten flour
30 g ungluten flour
10 g sugar
3 g baking powder
2 g active dried yeast
1 tsp oil
100 ml warm water

Method

1. Clean Chinese sausage with hot water, steam for a few minutes until cooked, cut into 2-3 pieces, brush with Hoi Sin sauce. Put aside for later use.
2. Sieve low gluten flour, ungluten flour, baking powder and sugar together, add oil and active yeast, mix well. Add warm water to bind and knead into a smooth dough.
3. On a floured board, knead the dough thoroughly (for about 5-8 minutes) until smooth and non-sticky.
4. Put dough on greaseproof paper, proof dough in a warm enclosed area for about 30 minutes.
5. Take out dough, press to release the gas, fold and knead slightly to form a very smooth and non-sticky dough. Divide into 4-6 portions, roll each into a long strips, coil in a piece of Chinese sausage. Position well on a small piece of greaseproof paper. Stand aside to proof for another 15 minutes. Pictures 1-3
6. Steam Chinese Sausage Buns over high heat for 10 minutes. Serve hot.

Practical Tips

- Add yeast into warm water with a little sugar until foamy (if active yeast is not used) and mix into floury ingredients.
- Observe the speed of rising during the first proof of dough. The temperature and time will affect the softness of bread dough.

碗仔翅
"Shark's Fin" Soup

碗仔翅
"Shark's Fin" Soup

份量：8 碗 / Makes 8 bowls

材料

粉絲 40 克
瘦肉 75 克
冬菇 5 朵
薑 1 片

湯料

上湯 1 公升
糖、鹽、生抽、老抽、
麻油各 1 茶匙

芡汁

馬蹄粉 2 湯匙
水 4 湯匙

伴食

胡椒粉、浙醋各適量

做法

1. 粉絲浸軟，略為剪碎，瀝乾。
2. 瘦肉焓熟拆成幼絲；冬菇浸軟切絲。
3. 用 1 湯匙油起鑊，爆香薑片棄掉；灒酒加入湯料煮滾。
4. 將所有材料加入湯料內，煮滾埋馬蹄粉芡，推稠即可，享用時伴以胡椒粉及浙醋。

心得

- 埋芡時要用小火，否則馬蹄粉漿會和粉絲黏作一團。
- 粉絲當魚翅，乃家喻戶曉之美話。「碗仔翅」乃極受大眾歡迎之街頭小吃，不妨在家裏一試。

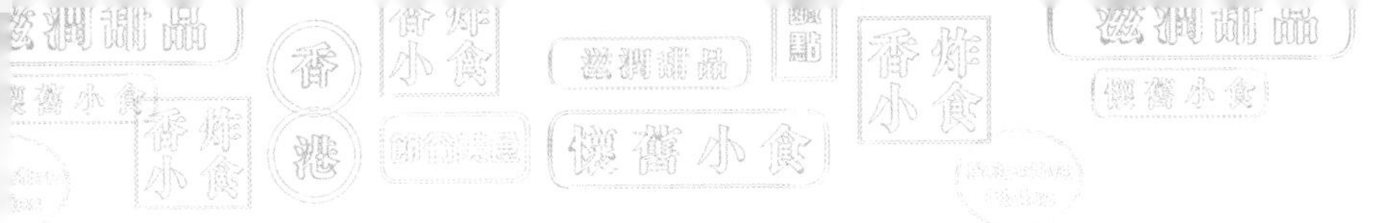

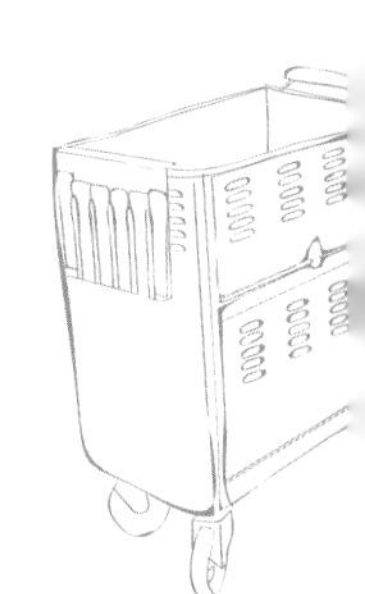

Ingredients

40 g mung bean thread
75 g lean pork
5 Chinese mushrooms
1 slice ginger

Ingredients for Soup

1 litre chicken stock
1 tsp sugar
1 tsp salt
1 tsp light soy sauce
1 tsp dark soy sauce
1 tsp sesame oil

Thickening

2 tbsp water chestnut flour
4 tbsp water

Serve With

pinch of pepper
pinch of red vinegar

Method

1. Soak mung bean thread until soft. Trim into lengths, drain.
2. Boil lean pork until cooked, tear into fine shreds; soak and shred Chinese mushrooms.
3. Heat 1 tbsp of oil in wok, sauté the ginger slice, discard. Sizzle in wine and add ingredients for soup, bring to the boil.
4. Add all shredded ingredients, mix in the thickening when re-boils. Serve hot with pepper and red vinegar.

Practical Tips

- Turn down the heat when adding in the thickening; otherwise the water chestnut flour cooks and turns into lumps easily in high heat.
- Mung bean thread looks like shark's fin when cooked and the soup is so named. It is a very popular snack in the stalls in Hong Kong Streets. Try it at home today.

生菜魚滑湯

Chinese Lettuce and Fish Strips Soup

這道街頭小吃，最讓人回味的是那股清香撲鼻的麻油味和帶點辛香的胡椒粉。

材料

上湯 500 毫升
中國生菜 1 棵
絞碎鯪魚肉 100 克
薑 1 片
麻油 1 茶匙

調味料

鹽 1/4 茶匙
胡椒粉少許
生粉 1 湯匙
蛋白 1 湯匙

做法

1. 中國生菜洗淨切絲。
2. 鯪魚肉加入調味料，用竹筷子順一方向攪拌至幼滑。
3. 燒滾上湯。
4. 燒水約 750 毫升，加入薑片，滾至 1-2 分鐘至出味；用刀將魚肉削成片狀直接入鍋，大火滾熟，撈起直接放入滾上湯內，撒入生菜絲及灑上麻油，拌勻後立即離火；熱食。隨個人喜好與胡椒粉同上。

心得

- 生菜魚滑湯的特色在於一股濃郁的麻油香味及魚肉（魚滑）幼滑的口感；更多的是一絲懷舊的情意。

Ingredients

500 ml chicken stock
1 stalk Chinese lettuce
100 g minced dace
1 slice ginger
1 tsp sesame oil

Seasonings

1/4 tsp salt
shakes of pepper
1 tbsp cornstarch
1 tbsp egg white

Method

1. Clean and cut Chinese lettuce into shreds.
2. Add seasonings into minced dace, stir with a pair of bamboo chopsticks in one direction until fluffy and smooth.
3. Bring chicken stock to the boil.
4. Boil about 750 ml water, add ginger, boil for 1-2 minutes. Using a knife, splash in minced fish in strips form, boil fish strips till cooked, drain into the boiling stock (with a ladle), add shredded lettuce and sesame oil, turn off heat immediately. Dish and serve hot with a good shake of pepper as desired.

Practical Tips

- The distinctive flavour of the soup comes from the sesame oil and the silvery smooth texture of the fish strips. It is also a typical street-stall oldies.

粉仔牛肉粥
Congee with Minced Beef

份量：6 碗 / Makes 6 bowls

材料

米 100 克
鹽 1.5 茶匙
水 2 公升
絞碎牛肉 100 克
銀絲米粉 25 克

調味料

蛋白 2 茶匙
水 2 茶匙
粟粉 1 茶匙
油 1 茶匙
鹽 1/4 茶匙
胡椒粉、麻油各少許

Ingredients

100 g rice
1.5 tsp salt
2 litres water
100 g minced beef
25 g rice vermicelli

Seasonings

2 egg whites
2 tsp water
1 tsp cornstarch
1 tsp oil
1/4 tsp salt
shakes of pepper
dash of sesame oil

做法

1. 牛肉加調味料拌勻。
2. 米粉剪碎，置滾油內炸鬆，撈起加入牛肉 圖 1~2，用手黏合，置碗底內。
3. 白米洗淨加鹽略醃，加水 2 公升煮滾，去泡，收慢火煮 1.5 小時。
4. 將大滾之白粥倒入盛有牛肉的碗內，攪至牛肉變熟即可。

心得

- 炸米粉時可先用少許米粉測試油溫，當米粉置油中立刻迅速浮起，表示油已夠熱。
- 宜用手撈勻牛肉及炸米粉，如嫌不夠衛生可用膠手套。若怕牛肉不熟，可將牛肉放入煮好的粥內滾熟，但切忌過火。

Method

1. Season the minced beef and mix well.
2. Trim rice vermicelli into tiny bits. Sprinkle into a pot of hot oil, dish and drain immediately. Mix with minced beef Pictures 1~2, leave in deep bowls.
3. Rinse and marinate the rice with salt. Add 2 litres of boiling water, boil and simmer for 1.5 hours.
4. Pour boiling congee into beef mixture. Mix and serve when beef turns colour.

Practical Tips

- Drop in bit of rice vermicelli to test the readiness of oil. Rice vermicelli should rise and float immediately if the heat is suitable.
- Traditionally, minced beef is rubbed into the fried rice vermicelli with fingers. To be hygienic, polythene gloves can be used. And to cook beef thoroughly, it can be added into boiling congee, but do not overcook the beef.

艇仔粥
Congee with Assorted Fish

份量：6 碗 / Makes 6 bowls

材料

米 75 克
水 2 公升
鹽 1 茶匙
水發魷魚 100 克
炸魚片 75 克
水發豬皮 75 克
胡椒粉、鹽、糖各少許

伴食

葱絲及炸花生

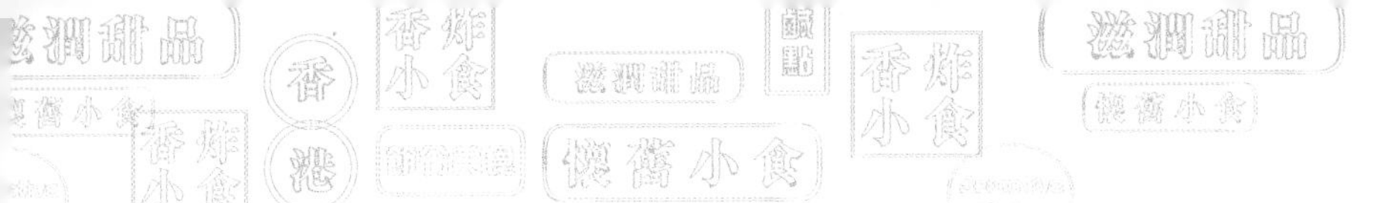

做法

1. 將水發魷魚、炸魚片及水發豬皮切絲。
2. 將米洗淨，加鹽 1 茶匙略醃，加水 2 公升煮滾，收慢火煮 1.5 小時。
3. 魷魚及豬皮絲放滾水內洗淨，瀝乾，與炸魚片絲一同放入煲好之白粥內，加胡椒粉、鹽、糖各少許，拌勻。
4. 灑上葱絲及花生即可供吃。

心得

- 水發魷魚及豬皮一般在凍肉公司及豆腐店有售。切絲後要用熱水沖去鹼水味才可使用。

Ingredients

75 g rice
2 litres water
1 tsp salt
100 g blanched squid
75 g fried fish cube
75 g blanched pig's skin
shakes of pepper
pinch of salt and sugar

To Serve

shredded spring onion and roasted peanuts

Method

1. Shred the blanched squid, fried fish cube and blanched pig's skin.
2. Rinse and marinate the rice with salt. Add in 2 litres boiling water. Simmer for 1.5 hours.
3. Rinse shredded squid and pig's skin in boiling water, drain and add in congee together with shred ingredients. Season with pinch of pepper, salt and sugar. Mix well.
4. Sprinkle shredded spring onion and roasted peanuts on top.

Practical Tips

- Blanched squid and pig's skin are sold in most frozen meat shops and bean curd shops in wet markets. Rinse shredded squid and pig's skin thoroughly in hot water before use.

糖不甩
Mini Sugary Dumplings

份量：20 粒 / Makes 20 dumplings

材料

糯米粉 275 克
暖水 375 毫升
壓碎烘脆花生 100 克
砂糖 75 克

糖水

片糖 100 克，切碎
水 125 毫升

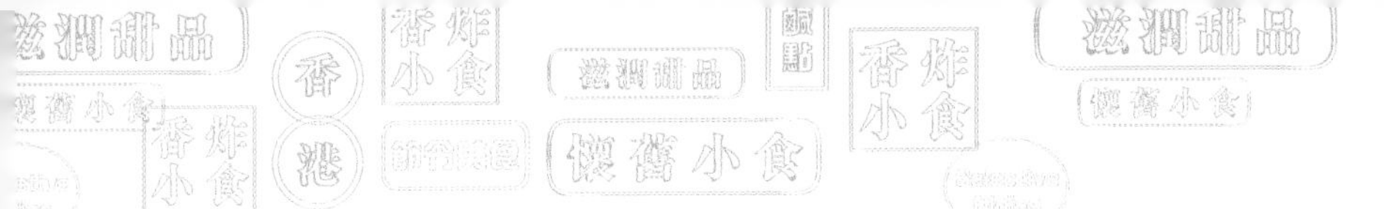

做法

1. 將 125 毫升水煮滾，加入片糖碎，慢火煮成糖水。
2. 糯米粉用適量暖水開勻，搓成一軟粉糰，再分搓成小粒，置大滾水內煮至浮起；撈起，再放入片糖水內，慢火煮片刻，隔去糖水置碟上。
3. 趁熱灑上花生碎及砂糖，即可供吃。

心得

- 可將整件搓好的粉糰用大火隔水蒸至透明及熟透，再置煲內保溫，食時才因應所需份量剪成小塊，趁熱灑上砂糖及花生碎。

Ingredients

275 g glutinous rice flour
375 ml warm water
100 g roasted peanuts, crushed
75 g castor sugar

Syrup

100 g slab sugar, crushed
125 ml water

Method

1. Prepare syrup: Boil 125 ml water, add in the crushed slab sugar, slow boil until dissolved.
2. Add sufficient warm water to glutinous rice flour, mix and knead to a soft dough, roll and divide into small portions, shape into mini dumplings, cook in fast-boiling water till floating on top. Remove and put into syrup, simmer for a short while, drain and dish.
3. Sprinkle crushed peanuts and castor sugar on top of hot sugary dumplings to serve.

Practical Tips

- The dough can be steamed over high heat until cooked. Keep warm in steamer. To serve, cut into small pieces and coat with castor sugar and crushed peanuts.

合桃酥
Walnut Cookies

份量：20 件 / Makes 20 pieces

材料

麵粉 150 克
梳打粉 1/4 茶匙
雞蛋 1 個
豬油 75 克
軟黃砂糖 80 克
蛋黃（塗餅用）1 個

做法

1. 預熱焗爐至攝氏 200 度，焗盤掃油。
2. 麵粉及梳打粉同篩勻，中間開穴，加入雞蛋、豬油及軟黃砂糖，拌勻，搓成光滑粉糰（不宜搓得太久），分成 20 小粒。
3. 將每份粉糰搓圓按扁，塗上蛋黃，放焗盤上，入爐焗約 15 分鐘至金黃色，取出，稍為冷卻即可供吃。

心得

- 剛出爐之合桃酥較軟身，須待冷才呈鬆脆效果。

Ingredients

150 g plain flour
1/4 tsp bicarbonate of soda
1 egg
75 g lard
80 g soft brown sugar
1 egg yolk (to glaze)

Method

1. Preheat oven to 200°C, grease the baking pan.
2. Sift plain flour and bicarbonate of soda together, make a well in the centre, add egg, lard and soft brown sugar, knead to a smooth dough (do not over knead), divide into 20 small balls.
3. Roll and press each piece of dough, glaze with egg yolk, bake for 15 minutes until golden brown. Remove and cool for a while, serve.

Practical Tips

- Hot walnut cookies are soft, cool thoroughly to get a crispy result.

蛋撻
Egg Tarts

份量：6 個 / Makes 6 egg tarts

餅皮料

麵粉 100 克
糖霜 10 克
鮮奶 1 湯匙
牛油 40 克
豬油 10 克
雞蛋 1/2 個（拂勻）
鹽少許

蛋漿料

鮮奶 150 毫升
幼砂糖 20 克
雞蛋（大）1 個

做法

1. 預備蛋漿：大雞蛋與砂糖拌勻，逐少加入鮮奶，過濾成幼滑蛋漿，候用。
2. 預備餅皮：牛油及豬油拌勻，打至軟滑，加入糖霜，再打一會，逐少拌入雞蛋及奶；麵粉與鹽同篩勻，分次加入牛油漿內，拌成粉糰，搓按成軟滑餅皮，包好，置雪櫃內冷藏 15 分鐘。
3. 預熱焗爐至攝氏 200 度。
4. 取出餅皮，分成 6 份，撻模掃油，放入一份餅皮，用手指一邊轉動一邊推按至均勻地鋪滿撻模，倒入餡料至九成滿。圖 1~6
5. 將蛋撻置焗爐內焗 15 分鐘後，將焗爐調校至攝氏 180 度再焗 10 分鐘，取出，熱食。

心得

- 焗爐溫度要因應焗爐的大小來調校。
- 注入蛋漿前，可用小叉在餅皮上略為刺孔。圖 4
- 餅皮料內可不用豬油，改用 50 克牛油來代替。
- 餅皮不宜過分搓撻。

Pastry

100 g plain flour
10 g icing sugar
1 tbsp milk
40 g butter
10 g lard
1/2 beaten egg
pinch of salt

Filling

150 ml fresh milk
20 g castor sugar
1 egg (large)

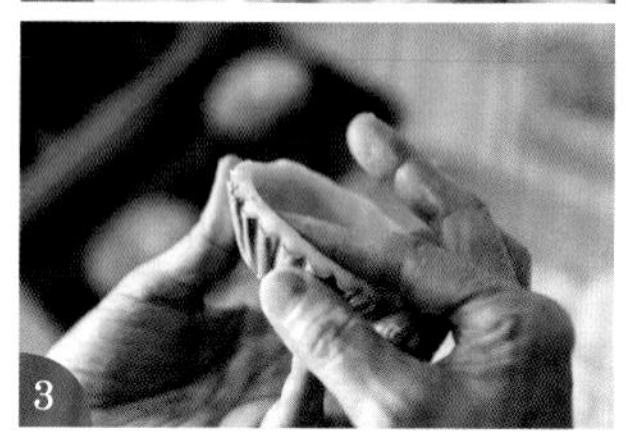

Method

1. Filling: Beat the large egg well with the castor sugar, add milk gradually, drain to get a smooth egg custard. Mix.
2. Pastry: Soften and beat butter and lard together, cream in icing sugar, add milk and beaten egg gradually, cream well between each addition. Sieve plain flour with the salt, gradually add into the buttered cream, mix slightly to form a smooth pastry. Wrap and chill for 15 minutes.
3. Preheat oven to 200°C.
4. Divide pastry into 6 portions, put each portion in a greased tart mould, press and turn to line the mould evenly just to the rim. Pour in egg custard to 90% full. Pictures 1~6
5. Bake for 15 minutes, lower the heat to 180°C and bake for another 10 minutes until set.

Practical Tips

- Adjust oven temperature according to the size of oven used.
- Can prick hole in pastry before adding in filling. Picture 4
- Use 50 g of butter to replace the use of lard.
- Avoid over kneading of buttered pastry.

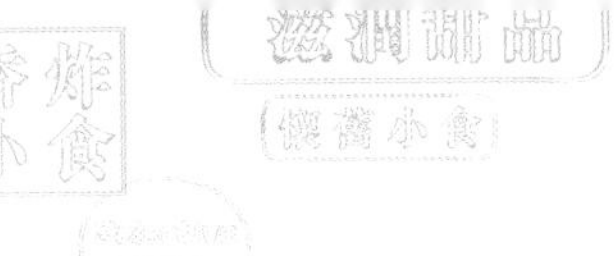

椰撻
Coconut Tarts

份量：6 個 / Makes 6 tarts

餅皮料

麵粉 100 克
牛油 50 克
冰水 1.5 湯匙
鹽少許

裝飾

車厘子 2 粒，切片

餡料

椰絲 50 克
糖 50 克
牛油 1 湯匙
淡奶 1 湯匙
發粉 1/4 茶匙
水 75 毫升
雞蛋 1 個（打匀）

做法

1. 預熱焗爐至攝氏 190 度，撻模掃油。
2. 預備餅皮：麵粉、鹽同篩匀，加入牛油，切碎，然後用手指將麵粉與牛油捏搓成麵包糠狀，加入適量冰水，拌匀按成軟滑不黏手之粉糰，包好置雪櫃內冷藏一會。
3. 預備餡料：將水煮熱，加入糖及牛油，煮溶，離火，加入椰絲及其他材料，拌匀。
4. 取出餅皮，案板灑少許粉，將餅皮擀至 3 毫米厚度，用餅模將撻皮切成六個圓塊，放入已掃油的撻模內用手捏好，中央放椰茸餡至七成滿，車厘子切片放面點綴。
5. 置焗爐內焗 25-30 分鐘至呈金黃色，即成；脫模後上碟。

心得

- 用冰水開皮，可提高撻皮鬆化及酥脆程度。

椰撻
Coconut Tarts

Pastry

100 g plain flour
50 g butter
1.5 tbsp chilled water
a little salt

Filling

50 g desiccated coconut
50 g castor sugar
1 tbsp butter
1 tbsp evaporated milk
1/4 tsp baking powder
75 ml water
1 egg, pbeaten

Decoration

2 glazed cherries, sliced

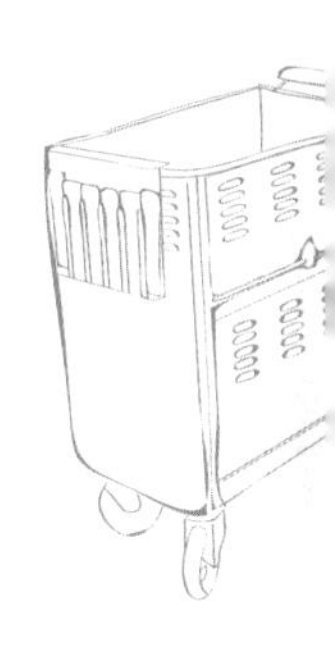

Method

1. Preheat oven to 190°C, grease tart cases.
2. Prepare pastry: sieve plain flour and salt together, add butter, cut into small pieces and rub with fingertips until mixture resembles fine breadcrumbs, add chilled water gradually, bind to form a smooth unsticky dough, wrap and chill dough for a while.
3. Prepare filling: heat water, add sugar and butter, stir until melted, remove from heat and add desiccated coconut, mix well and add in all other ingredients, mix well.
4. Take out pastry, roll to 3 mm thick on floured board, cut into 6 rounds with pastry cutter. Press and fit pastry into greased tart cases. Spoon in filling to 3/4 full, add sliced cherry to decorate.
5. Bake in preheated oven for 25-30 minutes until golden brown, unmould and serve hot.

Practical Tips

- Use chilled water for making the pastry helps to enhance shortness and crispiness of the tartlets.

雞蛋仔
Chinese Egg Puff

份量：3 底 / Makes 3 pieces

材料	Ingredients
麵粉 100 克	100 g plain flour
生粉 25 克	25 g tapioca starch
泡打粉 1 平茶匙	1 level tsp baking powder
雞蛋 2 個	2 eggs
砂糖 100 克	100 g castor sugar
淡奶 60 毫升	60 ml evaporated milk
清水 125 毫升	125 ml water

做法

1. 將麵粉、生粉及泡打粉篩勻候用。
2. 用木匙將雞蛋及砂糖拌勻，逐少加入淡奶及清水，拌勻後分次加入粉料中，不停攪拌成稀滑麵漿，不可起粒。
3. 將雞蛋仔模型底面兩面燒熱，掃油後注入粉漿至八分滿，蓋上蓋，將模夾緊，反轉，再置爐上，用中火底面各燒 1-2 分鐘至雞蛋仔離模及熟透。
4. 用叉將雞蛋仔挑出，趁熱進食。

心得

- 新購買回來之模型清洗乾淨後燒熱，然後按以上製法做出雞蛋仔。雞蛋仔取出後棄之不要，重複二至三次，直至雞蛋仔較易鬆離模型為止。

圖中的雞蛋仔模型，是當年走遍九龍新填地街、砵蘭街一帶選購的，現今已成為家中的一件珍藏。

Method

1. Sieve plain flour, tapioca starch and baking powder together.
2. Beat egg and sugar well with a wooden spoon, gradually mix in milk and water, pour into sieved ingredients, mix to a smooth, lump-free batter.
3. Heat the egg puff mould on both sides, grease it and pour in batter to 80% full, cover, grip and turn the mould upside down, cook over medium heat for 1-2 minutes on each side until set.
4. Unmould egg puff with a fork, serve immediately.

Practical Tips

- To treat a new iron mould, heat and grease after cleaning, prepare the batter, try 2-3 times, discard egg puffs. Repeat until mould becomes smooth and egg puffs no longer stick.

白糖沙翁
Sugary Egg Puff

白糖沙翁
Sugary Egg Puff

份量：24 件 / Makes 24 pieces

材料

清水 200 毫升
豬油 40 克
麵粉 80 克
雞蛋 3 個

蘸料

砂糖 1 量杯

Ingredients

200 ml water
40 g lard
80 g plain flour
3 eggs

Coating

1 cup castor sugar

做法

1. 燒滾清水，加入豬油待溶，離火，迅速篩入麵粉，不停攪動，以不起粉粒為合。圖 1~3
2. 立刻將雞蛋逐個拌入，攪成稠糊狀。圖 4~5
3. 燒油至暖，逐匙加入麵糊，慢火炸至蛋球脹大 3 倍及微爆口，瀝乾油分。圖 6~8
4. 趁熱蘸上砂糖即可。

心得

- 煮蛋糊時份量要準確，動作要迅速，蛋汁必須分次放入才較易將材料混合，記得離火。蛋球脹大只靠拌入之空氣，所以攪拌之動作很重要。油溫若過猛，會過早將蛋球之外皮硬化，空氣衝不出來，不能達至爆口效果，中心太軟難熟，所以火候也是炸蛋球關鍵之處。

Method

1. Bring water and lard to boil, remove from heat, sieve the flour, stir in sieved flour quickly, stir vigorously to avoid lumps. Pictures 1~3
2. Stir in eggs separately to form a thick paste. Pictures 4~5
3. Heat oil to lukewarm, scoop in flour mixture teaspoonful by teaspoonful. Deep-fry over gentle heat until egg puffs burst and expand to 3 times of the original size. Drain well. Pictures 6~8
4. Coat with castor sugar when still hot. Serve.

Practical Tips

- Ingredients should be weighed accurately. Remember to add flour away from heat, eggs should be stirred in separately and vigorously. The expanding of egg puff depends on the amount of air trapped in the mixture, so steps should be followed coherently. Avoid using high heat in deep-fat frying which will harden the surface of egg puffs to shortly preventing the air to burst for a fluffy result.

水晶餅
Crystal Cake

份量：16 件 / Makes 16 pieces

材料

澄麵粉 100 克
生粉 25 克
糖 75 克
豬油 2 茶匙
滾水 250 毫升
蓮蓉或豆沙 100 克

做法

1. 將澄麵粉及生粉混合，糖及豬油放滾水內煮溶，立刻沖入粉料中，拌至透明及呈糰狀，加蓋焗片刻。
2. 倒出粉糰，放枱上搓至軟滑，分成 16 小粒，搓圓按扁成水晶皮。
3. 蓮蓉或豆沙分成 16 小粒，放在水晶皮上，收口搓圓按扁，放餅模內壓實成形後倒出，置掃上油的碟或蒸籠內，大火蒸 7 分鐘。
4. 趁熱掃上熟油，即可享用。

心得

- 水晶皮擺放太久會變硬，所以必須趁熱使用。

Ingredients

100 g ungluten flour
25 g tapioca starch
75 g castor sugar
2 tsp lard
250 ml boiling water
100 g red bean paste or lotus seed paste

Method

1. Mix ungluten flour and tapioca starch together. Bring sugar and lard to a boil, pour into flour mixture, stir immediately to cook the mixture and form a lump of dough. Cover for 1-2 minutes.
2. Knead dough until soft and smooth, divide into 16 small lumps, press into round crystal pastries.
3. Divide red bean paste or lotus seed paste into 16 small portions. Wrap into each piece of pastry, seal, shape and flatten. Mould to form patterns. Unmould and steam on a greased plate or a bamboo steamer with high heat for 7 minutes.
4. Glaze with cooked oil. Serve hot.

Practical Tips

- Knead and shape dough when it is still soft and hot.

缽仔糕
Clay-Pot Puddings

份量：20 小碗 / Makes 20 puddings

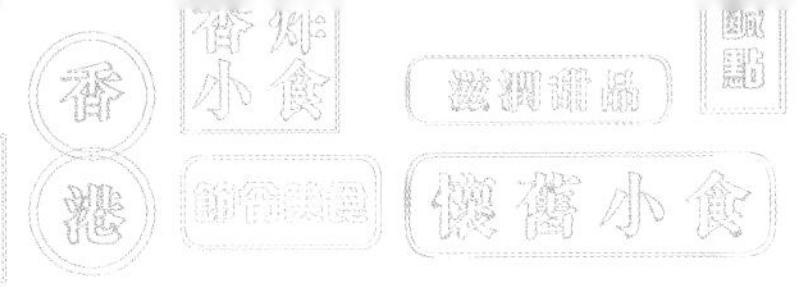

材料

粘米粉 100 克
糯米粉 1/2 湯匙
澄麵粉 75 克
砂糖（或片糖）100 克
水 500 毫升
浸透紅豆 3 湯匙
油少量
竹籤數枝

做法

1. 紅豆置滾水內用慢火煮軟，瀝乾候用。
2. 將粉料篩勻，用約 125 毫升水開勻成稠糊。
3. 將餘下之 375 毫升水煮滾，加入砂糖或片糖煮溶，趁熱撞入粉漿中，攪成滑粉漿。
4. 缽仔掃油，將粉漿倒入缽仔內至九成滿，加入紅豆，用大火蒸 20 分鐘後取出，稍候片刻才用竹籤挑出即可。

心得

- 熱糖水撞入粉漿時，需不停用木杓攪勻成幼滑粉漿。

圖中的小泥砵，是家翁當年回家鄉開平市趁墟期時為我採購的，感謝老人家的美意。日常生活中的人、物、事件，點滴在心頭，至今難忘。

Ingredients

100 g rice flour
1/2 tbsp glutinous rice flour
75 g ungluten flour
100 g castor sugar or slab sugar
500 ml water
3 tbsp soaked red beans
a little oil
a few bamboo skewers

Method

1. Simmer red beans in sufficient boiling water till soft and tender. Drain.
2. Sieve powdery ingredients, mix in 125 ml water to form a thick paste.
3. Boil the remaining 375 ml water, add sugar to form syrup, pour into flour mixture, stir well to form a smooth batter.
4. Grease clay-pots, spoon in batter to nearly full, steam over high heat for about 20 minutes or until set. Unmould with bamboo skewers after several minutes. Serve hot.

Practical Tips

- Keep stirring the flour mixture when pouring in the hot syrup.

白糖糕
White Sugar Sponge

份量：20 件 / Makes 20 pieces

材料

粘米粉 350 克
澄麵粉 100 克
清水 1 公升
蛋白 1/2 個
白糖 600 克

糕種

温開水 125 毫升
乾酵母 2 茶匙
糖 1/2 茶匙
粘米粉 1/2 杯

Ingredients

350 g rice flour
100 g ungluten flour
1 litre water
1/2 egg white
600 g castor sugar

Yeast Dough

125 ml warm water
2 tsp dry yeast
1/2 tsp castor sugar
1/2 cup rice flour

做法

1. 將糕種的乾酵母和糖用温開水浸至溶解，加入粘米粉 1/2 杯拌勻後，用濕毛巾蓋着，置室温 5 至 6 小時，待發起至比原來體積大兩至三倍。
2. 將粘米粉及澄麵粉用清水 500 毫升調勻，靜置至清水與濕粉分隔，將上層之水濾走。
3. 白糖加清水 500 毫升煮溶，待涼後加入蛋白拌勻，過濾，逐少倒入粉料中攪勻，蓋密待發酵數小時。
4. 拌入糕種搓勻，用濕布蓋盒口，靜置 12 小時發酵至呈細泡狀。
5. 將白糖糕漿傾入已鋪蒸糕布（預先塗油）之蒸籠內，隔水大火蒸 25 分鐘。
6. 倒出稍冷片刻，切件即成。

心得

- 發糕種的時間需視乎天氣而定，冬天寒冷需時較長，8-10 小時不等；夏天炎熱，需時較短，6-8 小時即可。

Method

1. Dissolve the dry yeast and 1/2 tsp of sugar in warm water, add 1/2 cup rice flour and mix to a soft dough. Cover with a wet towel and prove for 5-6 hours until the dough has expanded to 2 to 3 times of its original size.
2. Mix rice flour and ungluten flour with 500 ml of water. Set aside until water is separated from the wet flour mixture, filter water off the surface.
3. Dissolve sugar in 500 ml of boiling water, set aside to cool, add egg white, mix well, filter to get clear syrup. Pour syrup into wet flour mixture gradually, stir well, cover and prove the mixture for several hours.
4. Mix in yeast dough, knead slightly. Cover with a wet towel and prove for 12 hours until mixture raised like a bubbling sponge.
5. Line bamboo steamer with a piece of greased muslin, pour in the sugary mixture, steam over high heat for 25 minutes.
6. Cool slightly, cut into pieces and serve.

Practical Tips

- Time for proving yeast dough varied from season to season: in cold winter, it takes about 8-10 hours; in hot summer, it takes about 6-8 hours to rise.

芝麻卷
Black Sesame Seed Rolls

份量：10 件 / Makes 10 pieces

材料

粘米粉 50 克
馬蹄粉 50 克
澄麵粉 10 克
清水 500 毫升
黑芝麻 80 克
冰糖 75 克
麻油 1 茶匙

Ingredients

50 g rice flour
50 g water chestunt flour
10 g ungluten flour
500 ml water
80 g black sesame seed
75 g rock sugar
1 tsp sesame oil

做法

1. 粘米粉、馬蹄粉及澄麵粉拌勻，置深碗內候用。
2. 黑芝麻洗淨吹乾，用白鑊炒香，加 250 毫升水，置攪拌機內，打成芝麻漿，過濾隔去渣候用。
3. 冰糖用250毫升水煮成糖水，攤凍，與芝麻漿一同加入粉料內，拌勻，浸片刻，隔篩過濾，加入麻油。圖 1~2
4. 長方形蒸盤（約 14x20 厘米）掃油，平放蒸架上成水平線，蒸鑊內燒水至滾，注入粉料至填滿蒸盤（約 4 毫米厚），大火蒸 4 分鐘，揭蓋，待冷片刻。圖 3
5. 砧板上塗油少許，小心取出凝固了的芝麻片，打直切成 2 長條，向前推捲成芝麻卷。每底可做 2 件；重複直至完成餘下的芝麻粉漿為止，共 5 底，做成 10 件芝麻卷，雪凍，隨時上碟。圖 4~6

心得

- 若使用現成磨碎的熟芝麻粉代替生的黑芝麻，可用 200 毫升的水將 100 克的芝麻粉混和過濾後，作粉漿使用。但自磨芝麻漿的成品會較香。
- 蒸盤每次使用前要清潔乾淨及重新塗油。

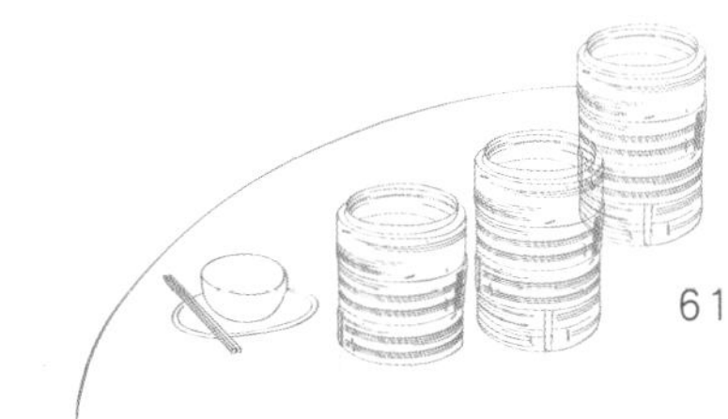

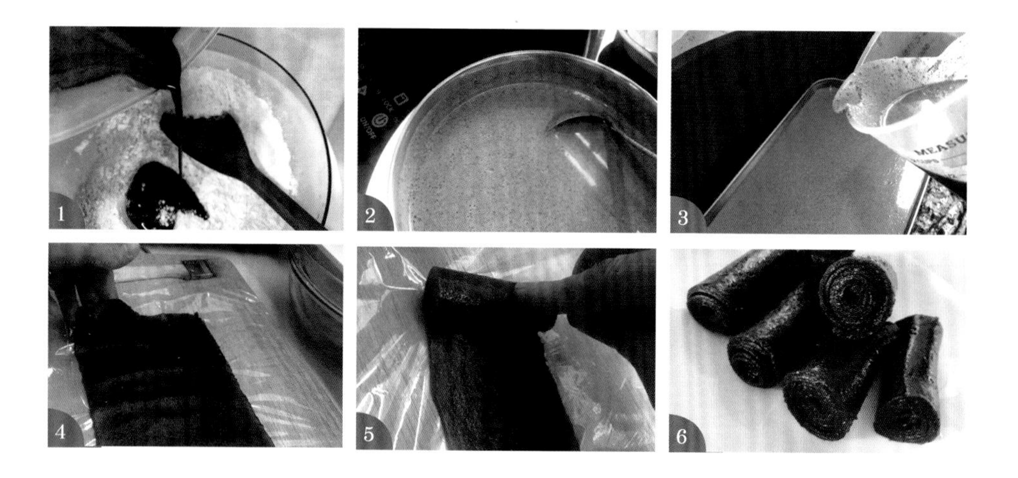

Method

1. Mix powdery ingredients in mixing bowl.
2. Rinse and air dry black sesame seed, stir fry in pan till fragrant, ground in electric blender with 250 ml water, pass through a sieve, drain well.
3. Dissolve rock sugar in 250 ml water, cool. Add into black sesame seed solution, mix and pour into the flour mixture to form a thin batter, stand for a while, pass through a sieve to remove residue, add seed sesame oil. Pictures 1~2
4. Put a greased rectangular tin (about 14x20 cm) on the steaming rack in a steamer horizontally, heat water in steamer, pour in the sesame seed solution to just cover the base of tin, about 4 mm thick. Steam over high heat until set, cool down a bit. Picture 3
5. Remove the sesame sheet carefully onto a greased chopping board, cut into half lengthwisely to get a long strips. Roll to the end to form 2 sesame seed rolls. Repeat and finish with the rest of the solution to get 10 sesame seed rolls. Chill and serve. Pictures 4~6

Practical Tips

- If ready-made ground black sesame seed is used, mix 200 ml water to 100 g of ground sesame seed which as homemade ground sesame seed will be more fragrant.
- The tin must be well cleaned and greased every time before use.

雙色芝麻糕
Layered Sesame Pudding

材料

粘米粉 200 克
馬蹄粉 60 克
清水 900 毫升
砂糖 180 克
淡奶 125 毫升
黑芝麻 50 克
椰汁 125 毫升

Ingredients

200 g rice flour
60 g water chestnut flour
900 ml water
180 g castor sugar
125 ml evaporated milk
50 g black sesame seed
125 ml coconut milk

做法

1. 預備芝麻漿：黑芝麻洗淨，吹乾，用白鑊炒香，加 200 毫升水，置攪拌機內打成芝麻漿，過濾隔渣；取芝麻漿 125 毫升，候用。
2. 砂糖用 200 毫升水煮成糖水候用。
3. 粘米粉及馬蹄粉同置深碗內拌勻，加入 500 毫升清水，攪勻成粉漿，靜置片刻，加入糖水，隔篩過清雜質，加入淡奶，拌勻；平均分成兩份，一份拌入芝麻漿，另一份加入椰汁，成黑、白粉漿。
4. 方型糕盤掃油，置蒸籠內蒸熱，注入黑粉漿一層，大火蒸 3-4 分鐘，掀蓋，冷卻片刻，加入一層白粉漿，加蓋蒸熟，每層蒸 3-4 分鐘，重複至最後一層為黑粉漿，離火前全底糕蒸 10 分鐘。
5. 取出芝麻糕，冷卻片刻，離模後切成小塊，成雙色千層芝麻糕；冷熱吃皆可。

心得

- 每層粉漿入糕盤前要攪拌片刻及用湯杓加入黑、白粉漿來固定每層的厚度。
- 湯杓的大小要與糕盤的大小配合，可自由調控芝麻糕的厚度和層數。
- 注意要挑走氣泡才加上鑊蓋。

雙色芝麻糕
Layered Sesame Pudding

Method

1. Prepare sesame seed solution: clean black sesame seed, air dry, fry in pan till fragrant, ground in electric blender with 200 ml water, drain by passing the mixture through a sieve to get 125 ml sesame seed solution.
2. Dissolve sugar in 200 ml boiling water.
3. Put rice flour and water chestnut flour in mixing bowl, mix in 500 ml water, soak for a while, add syrup, drain well, mix in evaporated milk, divide into two equal portion, add coconut milk or sesame seed solution in each of the portions to form a black and a white runny batters.
4. Heat up a greased square tin in a steamer. Pour in a thin layer of black batter just covering the base. Steamed for 3-4 minutes until set, cool down a bit, add a second layer of white batter to the same thickness, steam for another 3-4 minutes. Repeat until finishing the top layer with the black solutions. Steam the whole pudding for 10 minutes before removing from the heat.
5. Cool down a bit, turn out the layered pudding on a chopping board, cut into small pieces, serve hot or cold.

Practical Tips

- Stir each of the batter well before putting to steam and use a ladle for measuring to control an even thickness of each layer.
- The size of the ladle helps to control the desired thickness and number of layers.
- Remove bubbles on top before covering the steamer.

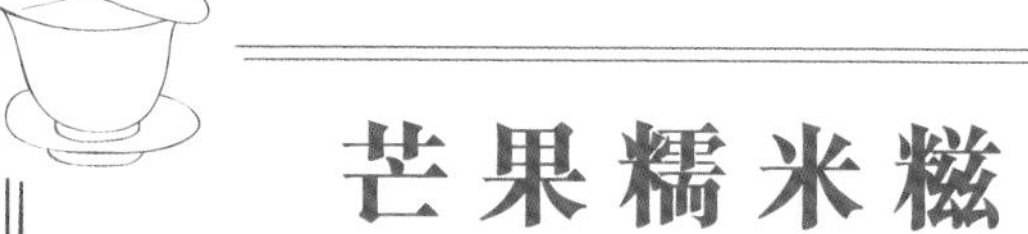

芒果糯米糍
Mango Glutinous Rice Dumplings

份量：10-12 粒 / Makes 10-12 dumplings

皮料

糯米粉 80 克
粟粉 20 克
砂糖 20 克
鮮芒果汁 100 毫升
鮮奶 75 毫升
油 1 茶匙

餡料

芒果 1 個

蘸料

椰絲 60 克

Dough

80 g glutinous rice flour
20 g cornstarch
20 g castor sugar
100 ml fresh mango juice
75 ml fresh milk
1 tsp oil

Filling

1 mango

Coating

60 g desiccated coconut

做法

1. 芒果去皮、去核，取肉，切粒（10-12 份）。
2. 糯米粉及粟粉篩勻，加入砂糖、芒果汁及鮮奶，拌勻成幼滑薄漿，加油，再拌勻。
3. 傾入已塗油的鋼碟內，大火蒸 20 分鐘至熟。看附圖
4. 將熟糯米糰倒在已塗油的保鮮紙上，待冷卻片刻。
5. 椰絲放平碟上，取糯米糰一大湯匙，放入椰絲中滾動一下，用手按扁包入芒果餡，埋口搓圓，再蘸上一層椰絲，再搓圓，上碟即成。

心得

- 匙羹一定要塗油才舀起熟糯米糰，可防止匙羹黏着熟粉糰。
- 包餡埋口時，盡量用手指尖操作。

Method

1. Peel and stone mango, take flesh, dice (10-12 cubes).
2. Sieve glutinous rice flour and cornstarch together, add sugar, fresh mango juice and fresh milk, stir and mix into a smooth runny batter, add oil, mix well.
3. Pour mixture into a greased metal plate, steam over a high heat for 20 minutes until cooked. As shown
4. Remove cooked mixture on a greased plastic wrap, cool down a bit.
5. Put desiccated coconut on a large plate, take a tablespoonful of cooked mixture, coat with a little desiccated coconut, shape flat with hands, wrap in a piece of mango cube, seal and roll into a ball shape, coat well with desiccated coconut. Reshape, dish and serve.

Practical Tips

- Handle the cooked mixture with a greased metal spoon to prevent sticking.
- Use fingertips only when shaping and sealing the dumplings.

香蕉糕
Banana Rolls

份量：12 件 / Makes 12 pieces

材料

糯米粉 125 克
糕粉 40 克
砂糖 75 克
暖水約 200 毫升
香蕉油 1 茶匙
豆沙 100 克

撒糕面粉料

糕粉 40 克

Ingredients

125 g glutinous rice flour
40 g pudding flour
75 g castor sugar
200 ml warm water
1 tsp banana essence
100 g red bean paste

For dredging

40 g pudding flour

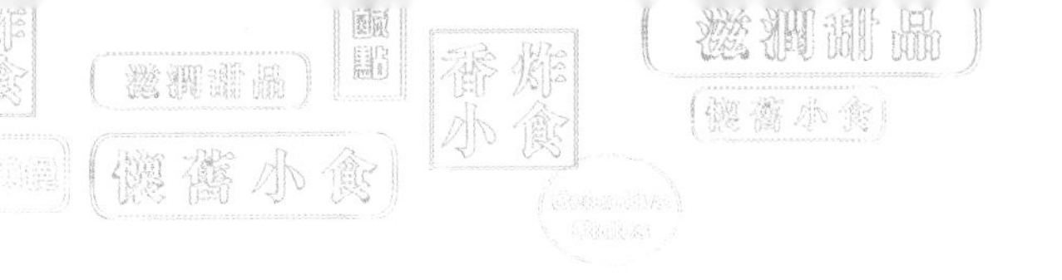

做法

1. 用白鑊將糯米粉炒熟（中火）至呈微金黃色即成糕粉。
2. 將糯米粉、糕粉及砂糖混合，加入香蕉油及適量暖水搓成軟粉糰，取其中半份用大火蒸 15 分鐘。
3. 將生熟皮混合，搓至極均勻，放枱上開成 1 厘米厚之長方塊。
4. 將豆沙置兩層牛油紙中間壓成薄塊，放在糯米皮上，推捲成條狀，置已塗油之碟上蒸 20-25 分鐘。
5. 取出香蕉糕，趁熱滾上糕粉，待冷片刻，用利刀切件即成。

心得

- 生熟皮要趁熱才較易搓勻，否則生熟皮會互不「相食」。
- 糕粉製法一：將糯米粉用白鑊慢火炒至微微出煙及呈淡黃色。
- 糕粉製法二：將糯米粉隔水蒸熟（中火），隔篩，即成糕粉。

Method

1. Stir-fry glutinous rice flour in a clean dry wok over medium heat until flour turns a light golden brown. It is the pudding flour.
2. Mix glutinous rice flour, pudding flour and sugar together, add banana essence and sufficient warm water to form a soft dough. Steam half of the dough over high heat for about 15 minutes.
3. Combine the two pieces of dough together by kneading vigorously.
4. Roll into a pastry of 1 cm thickness. Roll red bean paste to a thin sheet in between 2 layers of greaseproof paper, put on top of pastry. Roll and steam on greased plate for 20-25 minutes.
5. Coat banana roll with pudding flour when still hot. Cool and cut into pieces with a sharp knife. Serve.

Practical Tips

- To mix the dough, the steamed dough should be very hot; otherwise, it will not mix with the cold dough easily.
- To prepare pudding flour:

  Method 1: Stir fry glutinous rice flour over very low heat in a dry sauce pan (without oil) until slightly smoky and pale white in colour.
- Method 2: Steam glutinous rice flour over medium heat until cooked, pass through a sieve, use as required.

西米餅
Pearl Crystal Cakes

份量：20 件 / Makes 20 pieces

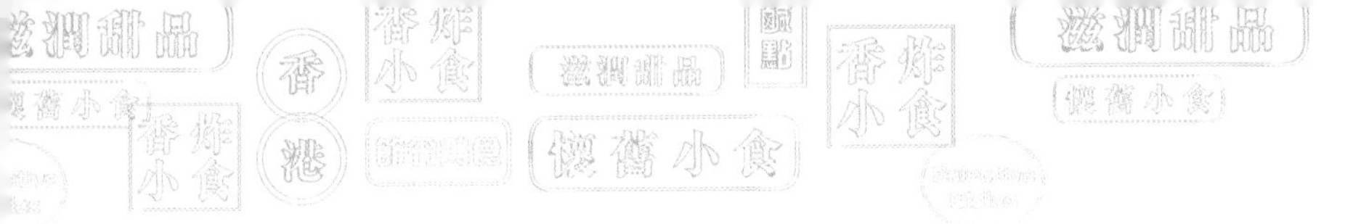

材料

西米 100 克
馬蹄粉 50 克
砂糖 30 克
清水 75 毫升
豆沙 50 克
油 1/2 湯匙

做法

1. 西米置深碗內，加入浸過西米面的滾水，加蓋浸焗 1 小時，置筲箕內沖水以去黏性，瀝乾候用。
2. 馬蹄粉加清水浸片刻，過濾入西米中，加砂糖及油拌勻。
3. 撻模掃一層油，加入 1 湯匙西米漿，放入豆沙一小粒，再蓋上 1 湯匙西米漿；用大火蒸約 8-10 分鐘至西米呈透明狀。
4. 離火，靜候片刻，挑出西米餅供食。

心得

- 西米須用大量滾水浸焗透徹，並間中攪拌，以免黏作一團。較大粒的西米可中途再換滾水浸焗。使用前將西米沖水瀝乾即可。

Ingredients

100 g sago
50 g water chestnut flour
30 g castor sugar
75 ml water
50 g red bean paste
1/2 tbsp oil

Method

1. Pour sufficient boiling water in a deep bowl of sago, cover and soak for 1 hour. Drain in colander, rinse under tap water, drain well.
2. Soak water chestnut flour in water for a while, sift into sago and mix in sugar and oil.
3. Grease tart moulds, add a tablespoonful of sago mixture, put into a small lump of red bean paste in each mould, top with another tablespoonful of sago mixture. Steam over high heat in 8-10 minutes or until sago turns to transparent.
4. Let cool, unmould and serve.

Practical Tips

- Sago should be soaked in boiling water, stir occasionally to prevent sticky. For larger grains, soak into another bowl of boiling water after half an hour. Rinse and drain well before use.

節令美食
Festival Specials

剛剛吃過五月糉，還有幾個月圓之夜才到中秋，媒體間的月餅廣告已百花齊放，互相爭妍。香港的節慶，此起彼落，是另一本土城市的生活寫照。

美食標誌人生，的確如此。每逢佳節，與其在坊間訂購一些包裝有點兒過分華麗的糕點，又或想避免吃下一些不知名的食物添加劑，最環保及健康不過的是自家製作自家品牌的應節食品了。香港人一向重視節慶美食，「年晚煎堆、人有我有」及喻意吉祥「步步高升」的各款賀年糕點，送禮自奉皆宜，還有端午節的糯米糉子，中秋節人月兩圓的雙黃蓮蓉月、為家中長者祝壽的壽桃，祝賀添丁之喜的薑蛋醋，於此篇章，一一奉上。應節食品當前，佳節好時年，你我互相記掛，家家聚首一堂說說笑笑，無論是司廚的、大快朵頤的，一如油鑊中的笑口棗，天天笑口常開，全年快樂安康！

鹹蛋散
Savoury Egg Bows

鹹蛋散
Savoury Egg Bows

材料

高筋麵粉 175 克
泡打粉 1/4 茶匙
梳打粉 1/8 茶匙
五香鹽 1/4 茶匙
南乳汁 2 湯匙
黑芝麻 2 湯匙
雞蛋 1 個（拂勻）
清水 80 毫升

Ingredients

175 g strong flour
1/4 tsp baking powder
1/8 tsp bicarbonate of soda
1/4 tsp spicy salt
2 tbsp red fermented bean curd juice
2 tbsp black sesame seed
1 beaten egg
80 ml water

做法

1. 將粉料一同過篩，加入五香鹽、南乳汁及黑芝麻，拌勻，倒入蛋汁及清水徐徐拌勻成糰，搓至不黏手為止，靜置 15 分鐘。
2. 用木棍將粉糰擀薄，切成 2.5×5 厘米的小方塊，在每一塊中央剎一小裂縫，將一端自開口處穿過，反轉成蝴蝶結。圖 1~6
3. 蛋散放入中火油鍋內，炸至鬆脆及呈金黃色後，撈起；將油溫加熱，把蛋散回鑊再炸一次，撈起，瀝乾油分，攤凍。便成另一賀年小吃。

心得

- 甜蛋散已在初版的《香港特色小吃》內介紹過了，只要用 1 湯匙糖代替南乳汁、五香鹽及黑芝麻便可，製法一樣，甜蛋散吃時可蘸少許糖漿。

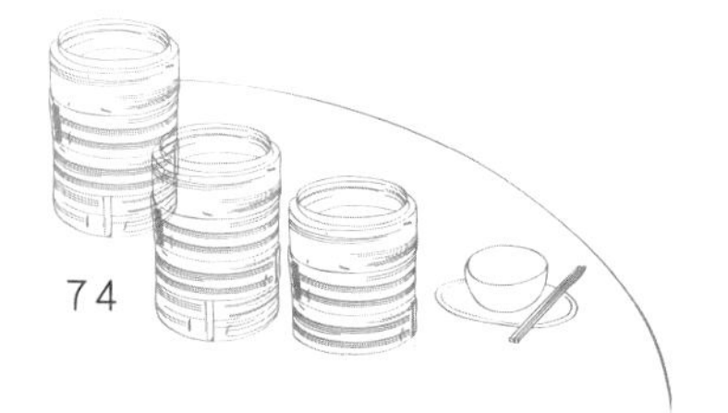

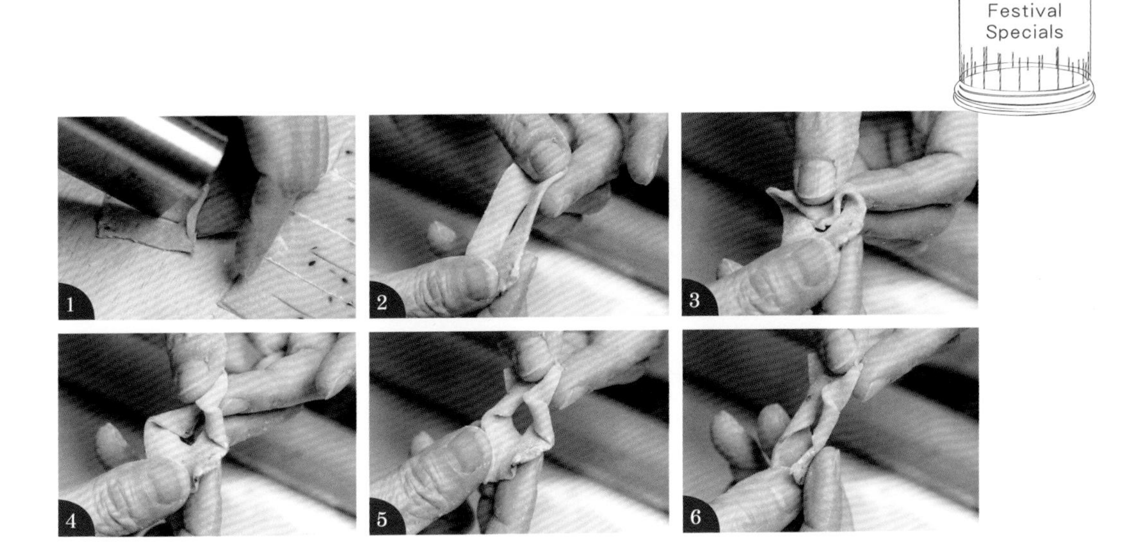

Method

1. Sieve all powdery ingredients together, add spicy salt, red fermented bean curd juice and black sesame seed, add beaten egg and water, mix well to form a soft dough, knead until smooth and not sticking. Set aside for 15 minutes.
2. Roll the dough into a thin sheet on a floured board, trim and cut into small rectangles of about 2.5x5 cm in size, make a sharp slit in the centre of each piece, pull one end through the slit, turn and fold to form a bow shape. Pictures 1~6
3. Deep-fry egg bows in medium high heat till golden brown and crispy. Drain for a while, increase temperature of oil, add in all the egg bows, deep-fry again for about 1 more minute. Dish, drain well, cool and store as another Chinese New Year Snacks.

Practical Tips

- To make Sweet Egg Bows, replace spicy salt, red fermented bean curd juice and black sesame seed with 1 tbsp of sugar, follow the same steps to get a change of taste. Sweet Egg Bows can be served with a golden syrup dip.

八寶糯米飯
Eight Treasures Rice

份量：4 碗 / Makes 4 bowls

材料

糯米 225 克
砂糖 75 克
豬油 25 克
豆沙 100 克
雜色提子乾 2 湯匙
紅車厘子（櫻桃）2 粒

Ingredients

225 g glutinous rice
75 g castor sugar
25 g lard
100 g red bean paste
2 tbsp raisins
2 red glazed cherries

做法

1. 將紅車厘子一開二。
2. 糯米洗淨浸過夜，隔水，大火蒸熟，趁熱加入砂糖及豬油拌勻。
3. 飯碗掃油，放入提子乾排成圖案，加約 4 湯匙甜糯米飯，按實，放入少許豆沙，再加甜糯米飯至滿，掃平。
4. 蓋上錫紙隔去倒汗水，大火蒸 10-15 分鐘。
5. 取出，反轉底向上即成。

心得

- 蒸糯米飯之時間需 1 小時或以上，如欠耐性，可用電飯煲煮糯米飯，比例是：2 杯糯米用 1.5 杯水，但蒸的糯米飯口感較佳。

Method

1. Cut glazed cherries into halves.
2. Rinse and soak glutinous rice overnight, drain and steam over high heat till cooked, mix in castor sugar and lard when still hot.
3. Grease rice bowls, line with raisins to form patterns, fill with 4 tbsp of sweetened glutinous rice, add a small piece of red bean paste in the centre. Top with another layer of rice to fill the bowls. Flatten the top.
4. Cover with foil and steam for 10-15 minutes over high heat.
5. Unmould with underside up. Serve.

Practical Tips

- Steamed glutinous rice takes about 1 hour or more to cook. For convenience, electric rice cooker can be used. Use 1.5 cups water to 2 cups of glutinous rice and cook as usual. Texture of steamed rice is better than boiled one.

甜湯丸
Dumplings in Thin Syrup

份量：20 粒 / Makes 20 dumplings

材料

糯米粉 175 克
澄麵粉 1.5 湯匙
暖水 175 毫升
片糖 1 片（切粒）
水 1 公升
片糖 150 克
薑 1 厚片

做法

1. 糯米粉與澄麵粉拌勻，加適量暖水搓成柔軟粉糰，分成 20 小粒，按扁，包入片糖粒，收口搓圓。
2. 燒熱水，加薑片煮至出味，加片糖煮至糖溶。
3. 湯丸放入大滾水內煮至浮起及微脹，撈起放入熱糖水內即可。

心得

- 片糖不要切得太大粒，否則湯丸熟後，片糖餡會仍未完全溶透。

Ingredients

175 g glutinous rice flour
1.5 tbsp ungluten flour
175 ml warm water
1 piece slab sugar, cut into cubes
1 litre water
150 g slab sugar
1 thick slice ginger

Method

1. Mix glutinous rice flour and ungluten flour together, add sufficient warm water to form a soft dough, divide into 20 small portions. Wrap in slab sugar, seal and roll into ball shape.
2. Bring water and sliced ginger to the boil, add slab sugar, cook until sugar is dissolved.
3. Cook dumplings in fast boiling water till dumplings float on surface. Drain into hot syrup. Serve.

Practical Tips

- Slab sugar cubes for filling should melt when dumplings are cooked, so the size of sugar cubes should not be too large.

笑口棗
Chinese Laughing Doughnuts

份量：24 個 / Makes 24 doughnuts

材料

麵粉 175 克
泡打粉 2 茶匙
砂糖 75 克
豬油 1 湯匙
冷開水約 75 毫升
白芝麻 1 杯

Ingredients

175 g plain flour
2 tsp baking powder
75 g castor sugar
1 tbsp lard
75 ml (approximately) cold water
1 cup white sesame seed

做法

1. 將麵粉及泡打粉篩勻，加入砂糖及豬油，用適量冷開水拌成一軟滑粉糰，靜置半小時，候用。
2. 將粉糰搓長，切成 24 小粒，搓圓，掃濕後再沾上一層白芝麻，搓圓。
3. 燒油至七成熱，離火，放入笑口棗，炸至浮面後開中火，炸至笑口棗爆裂及呈金黃色，盛起，瀝乾油分，待冷即可入罐。

心得

- 離火放入笑口棗，可避免芝麻炸得過火及可將笑口棗炸透；但不要太遲回火，否則笑口棗會吸油過多而變得肥膩。炸時要經常使用鑊鏟或炸籬轉動笑口棗。

Method

1. Sieve plain flour and baking powder together, add sugar and lard, mix in sufficient cold water to form a soft dough, set aside for half an hour before use.
2. Roll and divide dough into 24 small pieces, roll each piece of dough to a ball shape, damp with water, coat evenly with white sesame seed, shape well.
3. Heat oil until hot, turn off the heat, slide in doughnuts, return to heat when doughnuts start to float, deep-fry over medium heat until the doughnuts burst and turn to golden brown in colour, drain, cool and store.

Practical Tips

- Turning off the heat before sliding in the doughnuts can avoid sesame seed coating from over-burning. Cold oil tends to penetrate into doughnuts, so return to heat when doughnuts start to burst and keep turning the doughnuts with a ladle.

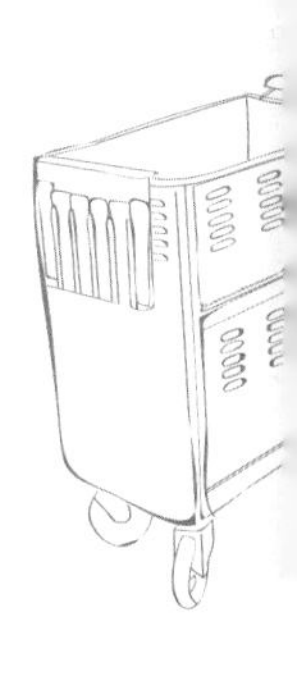

年糕
New Year Pudding

份量：1 底 / Makes 1 pudding

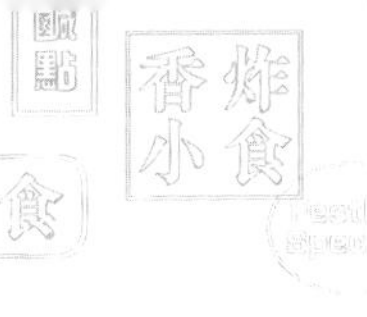

材料

糯米粉 180 克
澄麵粉 40 克
片糖 100 克（切碎）
暖水 250 毫升
杞子 2 湯匙
薑汁 2 湯匙
油 1/2 湯匙

做法

1. 杞子用暖水浸 30 分鐘，隔篩，將杞子水（250 毫升）煮滾，加入片糖碎，煮溶。
2. 糯米粉和澄麵粉篩勻，置深碗內，加入杞子糖水，輕手拌勻成掛杓麵漿 看附圖，拌入薑汁及油。
3. 糕盆掃油，注入麵漿，用中火蒸 1 小時至熟透，取出，用杞子點綴，即成。

心得

- 蒸糕途中蒸鑊內只可加入滾水，凍水會令水温下降。離火前可用竹籤測試年糕是否熟透，竹籤插入取出時，黏着竹籤的粉漿應呈透明狀，代表糕已熟透。
- 年糕可原底作賀年禮品，或切片煎香熱食。

Ingredients

180 g glutinous rice flour
40 g ungluten flour
100 g slab sugar, crushed
250 ml warm water
2 tbsp goji cherries
2 tbsp ginger juice
1/2 tbsp oil

Method

1. Soak goji cherries in warm water for 30 minutes, drain to get 250 ml goji cherry water, bring to the boil, add crushed slab sugar, heat and stir well until slab sugar is thoroughly dissolved.
2. Sieve glutinous rice flour and ungluten flour together. Pour in goji cherry solution, mix well to form a coating consistency As shown, add ginger juice and oil, mix well.
3. Grease a round pudding, tin pour in batter, steam over medium heat for 1 hour, decorate with goji cherry, cool and serve.

Practical Tips

- Add boiling water in steamer in between steaming. Test the readiness of pudding by piercing through the pudding with a bamboo skewer, the paste remain on skewer should appear sticky and transparent when cooked.
- New year pudding can be packed as a Chinese New Year gift, or cut into slices, shallow-fry and serve hot as a typical festive snack.

煎堆仔
Sesame Dumplings

份量：20 個 / Makes 20 dumplings

材料

糯米粉 275 克
水 375 毫升
片糖 125 克
白芝麻 250 克

Ingredients

275 g glutinous rice flour
375 ml water
125 g slab sugar
250 g white sesame seed

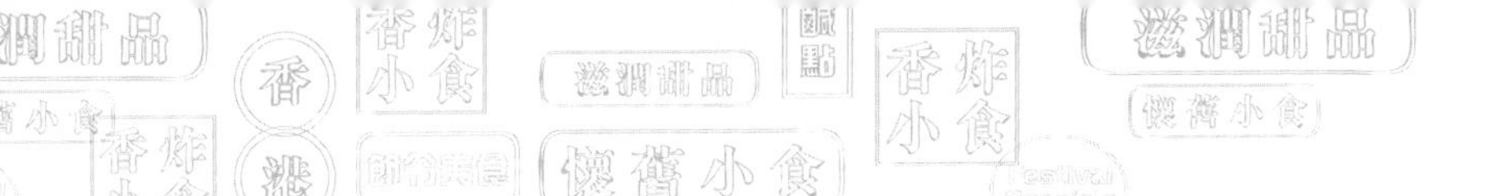

糯米粉受熱會釋出氣體，用煎堆鏟或平杓輕壓浮起的煎堆，讓氣體充分自由地使煎堆膨脹，令其呈皮薄通透效果。

做法

1. 片糖加水煮溶，攤至稍暖，加入糯米粉攪拌成糰，搓軟，分成 20 小粒，搓圓。
2. 掌心濕水，將糯米粉糰略搓，沾上白芝麻，搓圓。
3. 將油燒至七成熱，離火，放入煎堆，炸至略浮，開火，用平杓略壓煎堆使其逐漸脹大，並呈空心狀。見附圖
4. 炸透至皮脆及呈金黃色，瀝乾油分熱食。

心得

- 炸煎堆宜用大鑊油，讓煎堆有空間膨脹。壓煎堆之力要均勻及分次加壓，油溫不宜過猛，因芝麻很易上色。

Method

1. Dissolve slab sugar in 375 ml water. Cool down a bit. Mix with glutinous rice flour and knead to form a soft dough. Roll and divide into 20 small portions.
2. Damp palms a bit, roll each dumpling into a round, coat with white sesame seed, shape well.
3. Heat oil to 70% hot, turn off the heat, slide in dumplings, turn on the heat again when the dumplings start floating. Press them constantly with a ladle till the dumplings expand and become hollow in shape. As shown
4. Deep-fry sesame dumplings become crispy and turn to golden brown in colour. Drain well and serve hot.

Practical Tips

- Oil used should be sufficient to allow room for dumplings to expand. Apply pressure gradually when dumplings start to float and expand. Avoid using high heat because white sesame seed will get burnt easily.

豆沙角
Fried Red Bean Dumplings

份量：12 個 / Makes 12 pieces

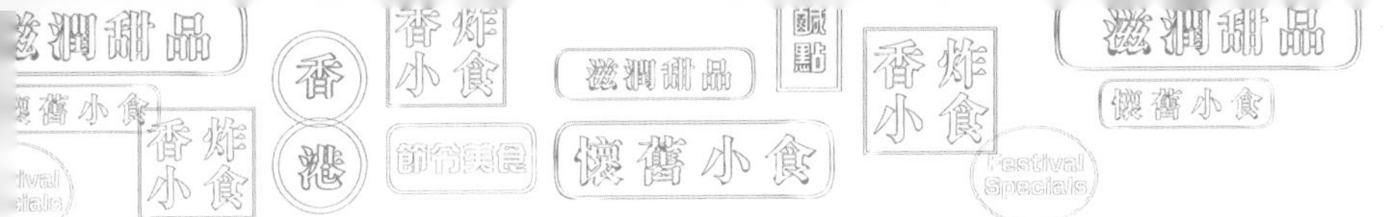

材料

糯米粉 275 克
澄麵粉 2 湯匙
暖水 375 毫升
砂糖 1 湯匙
豆沙 225 克
黑芝麻適量

Ingredients

275 g glutinous rice flour
2 tbsp ungluten flour
375 ml warm water
1 tbsp castor sugar
225 g red bean paste
black sesame seed

做法

1. 糯米粉與澄麵粉篩勻，加入適量暖水開成軟粉糰，分成兩份。
2. 將半份粉糰放在塗了油的碟上，蒸 10-15 分鐘至熟，趁熱與餘下之生粉糰搓勻，加入砂糖，搓至極均勻，搓長切成 12 份，壓成圓塊。
3. 豆沙搓長，分成 12 小粒，包入糯米皮中，做成三角形，蘸上黑芝麻，用大火油炸至呈微黃色即可。

心得

- 此半生熟皮料亦適用於製作鹹水角，但肉餡須預先炒熟，待涼後才用。搓皮時可加入少許豬油，使皮更鬆脆。

Method

1. Sieve glutinous rice flour and ungluten flour together, mix in sufficient warm water to form a soft dough. Halve into two.
2. Put half of the dough on a greased plate and steam for about 10-15 minutes until cooked. Mix steamed dough with the reserved dough, knead with castor sugar at the same time. Divide into 12 portions, press into round pastries.
3. Divide red bean paste into 12 small lumps, wrap each into pastries, shape into triangles, seal and decorate with some black sesame seed. Deep-fry until slightly brown in colour. Drain well and serve.

Practical Tips

- The dough is also suitable for making savoury meat dumplings, but fillings should be stir-fried and cooled before use. For a crispier result, knead in a lump of lard to the dough.

油角仔
Crispy Dumplings

份量：32 隻 / Makes 32 dumplings

皮料

麵粉 225 克
豬油粒 25 克
油 25 克
砂糖 50 克
蛋液 75 毫升

餡料

炒香白芝麻 50 克
壓碎熟花生 50 克
砂糖 75 克

做法

1. 將 175 克麵粉過篩，加入豬油粒，用手指搓入麵粉中至極幼細，加入砂糖，拌勻，逐少加入生油及蛋液，拌勻，逐少撥入餘下 50 克麵粉，輕搓成一不黏手之粉糰。
2. 將餡料混合，待用。
3. 枱上灑上乾麵粉，用棍將麵皮擀至 1 分厚，用圓玻璃杯口鈒成小圓塊。見附圖
4. 包入適量餡料，對摺成半月形，收口鎖邊，用中火油炸透，瀝乾油分，待冷即可入罐，隨時享用。

心得

- 包角仔時，皮的邊緣盡量不要沾有糖餡，否則炸油角仔時開口會很易爆開的。

Dough

225 g plain flour
25 g diced lard
25 g oil
50 g castor sugar
75 ml beaten egg

Filling

50 g fried white sesame seed
50 g crushed roasted peanuts
75 g castor sugar

Method

1. Sieve 175 g of the plain flour, rub in lard, add sugar, bind with oil and beaten egg, mix well. Gradually add in remaining flour, slightly knead to form a soft dough.
2. Mix ingredients for filling. Set aside.
3. Sprinkle some flour on the table. Roll the pastry to about 1mm thick. Cut into small rounds with a glass or plain cutter. As shown
4. Wrap in sufficient filling on each piece of pastry, half-fold, press rim to seal dumplings, pinch to form patterns on curve side. Deep-fry the dumplings till crispy and golden brown in colour. Drain, cool and store.

Practical Tips

- Keep rim of pastry rounds free from sugary filling or dumplings will crack in oil when deep-frying.

賀年茶泡
Assorted New Year Crispies

材料	Ingredients
花生 100 克	100 g peanuts
芋頭半個	1/2 taro
番薯 1 個	1 sweet potato
腰果 100 克	100 g cashew nuts
幼鹽適量	some salt

做法

1. 花生浸滾水，去衣，吹乾。
2. 芋頭及番薯去皮，切成小薄片，用滾鹽水浸片刻，撈起鋪平風乾。看附圖
3. 腰果用鹽水焓片刻，取出吹乾。
4. 將各茶泡料分別放滾油內炸至金黃色，撈起瀝乾油分，趁熱灑上幼鹽，待冷入罐。

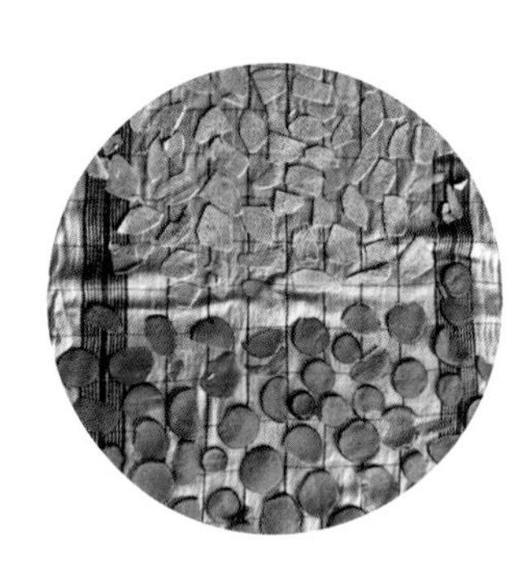

心得

- 各款茶泡料應獨立下油鑊，不要全部一起炸，因它們的受火程度不同。

Method

1. Soak peanuts in boiling water, skin and air-dry.
2. Peel taro and sweet potato, cut into small thin slices, rinse in hot salt solution, drain, spread and air-dry. As shown
3. Boil cashew nuts in salt water, drain and air-dry.
4. Deep-fry the above ingredients separately in hot oil until golden brown. Drain and dust with salt. Cool and store in air-tight tins.

Practical Tips

- Fry the assorted crispies separately, they brown at different temperature.

椰汁糖環
Coconut Rosettes

份量：80 件 / Makes 80 pieces

材料

麵粉 1 量杯
粟粉 1 量杯
砂糖 1 量杯
雞蛋 4 個
椰汁 175 毫升
淡奶 125 毫升
水 250 毫升
油適量（炸糖環用）

做法

1. 將麵粉、粟粉和砂糖混合，篩於深碗內，中間開一小穴，打入雞蛋，逐少加入椰汁、淡奶及水，不停攪拌成稀麵漿，靜置片刻。
2. 將油燒熱，放入糖環模加熱數分鐘後，將熱模浸入麵漿內（不可浸過模面）。圖 1
3. 立刻將模放入熱油中，炸至麵漿凝固及鬆脱，並呈金黃色。圖 2
4. 取出糖環，瀝乾油分，待冷後便可入罐，隨時供吃。

心得

- 糖環模必須燒至熱度適中才放入麵漿內，若模型不夠熱便上不到漿，若過熱，則未放入油，麵漿便已脱模。可將模型燒熱後，提起模型，讓多餘油分滴回鑊中，約數秒後立刻浸入麵漿，麵漿便上得恰到好處。

滋潤甜品
香炸小食
懷舊小食

節令美食
Festival
Specials

Metal mould should be well heated before dipping into batter.

Rosettes set and loosen in heated oil. Fry till crispy.

Ingredients

1 cup plain flour
1 cup cornstarch
1 cup castor sugar
4 eggs
175 ml coconut milk
125 ml evaporated milk
250 ml water
oil for deep-frying

Method

1. Sift plain flour, cornstarch and sugar in a deep bowl, make a well in the centre, drop in eggs, gradually pour in coconut milk, evaporated milk and water to form a thin, smooth batter; stand aside for a while.
2. Heat some oil, put in the rosette mould to heat for a few minutes. Dip the heated mould in batter just to the depth of the mould. Picture 1
3. Transfer the mould into hot oil, hold until the coated batter coagulates and separates from the mould. Picture 2
4. Drain the coconut rosettes when it turns to golden brown in colour. Cool, store and serve at anytime.

Practical Tips

- Heat the rosette mould just hot enough to coat the batter and not to cook the batter. Lift the heated mould, hold for a few seconds to drip excess oil, dip in batter immediately to form a nice coat.

蘿蔔糕
Turnip Puddings

蘿蔔糕
Turnip Puddings

份量：2 底 ／ Makes 2 pudding tins

材料

白蘿蔔 1.5 公斤
粘米粉 450 克
澄麵粉 75 克
冬菇 8 朵
蝦米 50 克
臘腸、臘肉 175 克
暖水 750 毫升
滾水 750 毫升

調味料

鹽 3 茶匙
糖 2 茶匙
麻油、胡椒粉各少許

裝飾

芫茜碎、炒香白芝麻各少許

做法

1. 冬菇浸軟，切粒；蝦米洗淨，浸軟；臘腸、臘肉洗淨後出水，蒸熟及切粒。
2. 粘米粉與澄麵粉篩入大碗內，加調味料，用 750 毫升暖水開成粉漿。
3. 冬菇、蝦米、臘肉及臘腸起油鑊，爆香，灒酒，炒勻盛起。
4. 白蘿蔔去衣刨絲，放油鑊略炒，加滾水 750 毫升，煮滾趁熱撞入粉漿內，拌成糊狀，加入其他材料拌勻。
5. 糕盤掃油，倒入糕料，大火蒸 1 小時或至糕熟。
6. 趁熱灑上白芝麻、芫茜碎及葱粒，待冷切件，用油煎香，趁熱供吃。

心得

- 吃蘿蔔糕的口味因人而異，如喜歡較硬的糕底，可多加少量粘米粉或減少蘿蔔也可。配料亦可隨意加減。

Ingredients

1.5 kg turnips
450 g rice flour
75 g ungluten flour
8 Chinese mushrooms
50 g dried shrimps
175 g Chinese sausage and preserved pork
750 ml warm water
750 ml boiling water

Seasonings

3 tsp salt
2 tsp sugar
dash of sesame oil
shakes of pepper

Garnish

dash of diced coriander and roasted white sesame seed.

Method

1. Soak and dice the Chinese mushrooms. Clean and soak the dried shrimps. Blanch and steam the Chinese sausage and preserved pork; dice.
2. Sieve rice flour and ungluten flour together into a large bowl, add seasonings, 750 ml of warm water to form a batter.
3. Sauté diced ingredients in oil, sizzle in wine, stir well. Dish.
4. Peel and grate turnips into long shreds, sauté with a little oil, add 750 ml of boiling water, bring turnips to boil, immediately pour mixture into the batter, stir to obtain a sticky consistency. Add other ingredients, mix well.
5. Pour pudding mixture into a greased tin. Steam over high heat for 1 hour until cooked.
6. Garnish, cool, slice and shallow-fry till it turns to golden brown in colour on both sides.

Practical Tips

- Texture of pudding can be adjusted according to one's desire. For a firmer texture, add a little more rice flour or use less turnips. Choices of assorted meat can be optional.

馬蹄糕
Water Chestnut Pudding

材料

馬蹄粉 250 克
水 1 公升
冰糖 300 克
馬蹄 10-12 顆
油 1 湯匙

Ingredients

250 g water chestnut flour
1 litre water
300 g rock sugar, crushed
10-12 pieces water chestnuts
1 tbsp oil

做法

1. 馬蹄粉置深碗內，加入 500 毫升水拌勻，浸片刻。
2. 冰糖放入餘下 500 毫升水內，用慢火煮溶，趁熱撞入馬蹄粉漿內，拌勻後過濾，置鍋內加油。
3. 馬蹄洗淨，去皮，切片或拍碎，加入鍋內，開火邊煮邊攪成粉糰及呈透明狀，迅速倒入已塗油的鐵模內，掃平，用錫紙蓋好，大火蒸 15-20 分鐘。
4. 離火，冷卻後切片，煎至呈金黃色，熱食。

心得

- 可隨意加少許吉士粉於馬蹄粉內，以增添顏色。

Method

1. Mix water chestnut flour with 500 ml water in mixing bowl, soak for a while.
2. Dissolve crushed rock sugar in the rest of 500 ml water under low heat, pour into flour solution, mix well, drain into saucepan, add oil.
3. Clean, peel, crush or slice water chestnuts, add into saucepan, cook the mixture until thickened and become transparent, keep stirring vigorously and pour into greased loaf tin, smoothen the top, cover with tin foil to avoid water vapour. Steam over high heat for 15-20 minutes.
4. Remove from heat, cool and slice. Shallow-fry sliced water chestnut puddings until golden brown on both sides. Serve hot.

Practical Tips

- A little custard powder can be added into water chestnut flour, for colour only.

芋頭糕
Taro Puddings

材料

芋頭粒 250 克
粘米粉 150 克
水 500 毫升
蝦米 20 克
冬菇 3 朵（浸軟）
臘腸 1 條
臘肉少許

調味料

鹽 1 茶匙
糖 1/2 茶匙
五香粉 1/4 茶匙
胡椒粉少許
麻油 1 茶匙
油 1 湯匙

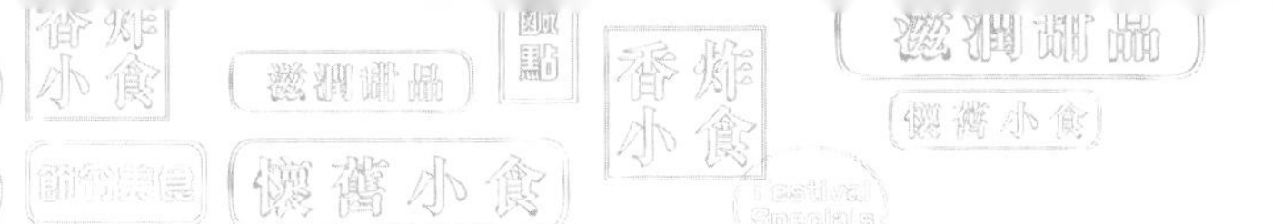

Ingredients

250 g taro, diced
150 g rice flour
500 ml water
20 g dried shrimps
3 soaked Chinese mushrooms
1 preserved sausage
a small piece preserved pork

Seasonings

1 tsp salt
1/2 tsp sugar
1/4 tsp five spice powder
pepper to taste
1 tsp sesame oil
1 tbsp oil

做法

1. 蝦米、冬菇、臘腸、臘肉分別切粒，用少許油炒香，盛起候用。
2. 粘米粉置深碗內，加入 200 毫升水，拌勻成幼滑粉漿。
3. 芋頭粒用餘下的 300 毫升水慢火煮 5-8 分鐘，離火趁熱注入粘米漿，加入調味料及炒熟的蝦米、冬菇、臘腸和臘肉，拌勻，攪勻成厚糊狀。
4. 趁熱傾入已塗油的鐵模內，掃平，大火蒸 30-35 分鐘至熟透，待冷，切件煎香，熱食。

心得

- 將粉漿注入芋頭內時必須離火，以防起粉粒，並要不停攪拌至成幼滑糊狀才拌入調味料及配料。可用竹籤測試芋頭糕是否熟透，竹籤插入取出時仍沾有白色粉漿即表示仍未熟透。

Method

1. Clean and dice, soaked dried shrimps, Chinese mushroom, preserved sausage and pork, stir fry in a little oil, dish.
2. Mix rice flour with 200 ml water in mixing bowl to form a smooth batter.
3. Cook taro in the rest of 300 ml water over low heat for 5-8 minutes until just cooked. Remove from heat and mix in rice flour solution, keep stirring to form a thick paste, stir in fried ingredients and seasonings, mix well.
4. Pour the hot thickened mixture into a greased tin, flatten the surface, steam the pudding over high heat for 30-35 minutes until cooked. Cool, slice and shallow-fry, serve hot.

Pratical Tips

- To test the readiness of taro pudding, insert a bambo skewer into the middle part, if it comes out clean and free from any sticky paste, the taro pudding is cooked through.

家鄉鹹肉糉
Rice Dumplings in Bamboo Leaves

份量：8 隻 / Makes 8 rice dumplings

材料

竹葉 24 塊
水草 8 條
糯米 600 克
鹽 1 茶匙
綠豆邊 175 克
冬菇 8 朵
腩肉 275 克
蝦米 40 克
瑤柱 4 粒
鹹蛋黃 4 個

腩肉醃料

鹽 1 茶匙
五香粉 1/2 茶匙
麻油少許

做法

1. 將竹葉浸透，剪去硬蒂；水草浸透；糯米洗淨後瀝乾，加鹽 1 茶匙拌勻；綠豆邊浸片刻，洗淨，瀝乾；冬菇浸軟後去蒂；腩肉洗淨，切件後用醃料拌勻待 3 小時；蝦米浸軟。瑤柱浸軟一開二，鹹蛋黃一開二。圖 1
2. 將兩塊竹葉重疊，屈摺成一碗形。圖 2
3. 底層放一大湯匙糯米和綠豆邊，放上各項餡料，最後加糯米。圖 3~5
4. 左手將糉兜托好，自前端另加竹葉一塊至二塊。圖 6
5. 將竹葉尾端豎起，左手將餡料封口，右手將末端竹葉向下屈摺，固定糉形。圖 7~8
6. 用水草將鹹肉糉紮實，置浸過面之滾水內烚 2 小時，撈起即成。圖 9~10

心得

- 為方便移動或撈起糉子，可用一粗水草將糉子串起才放進滾水內，或於水草末段結一小圈亦可。圖 11

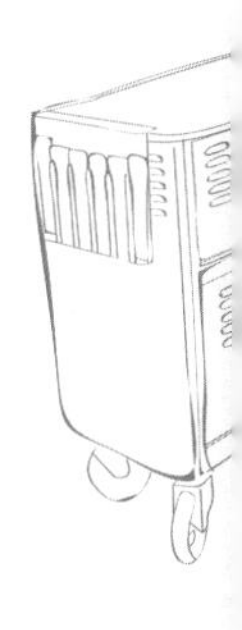

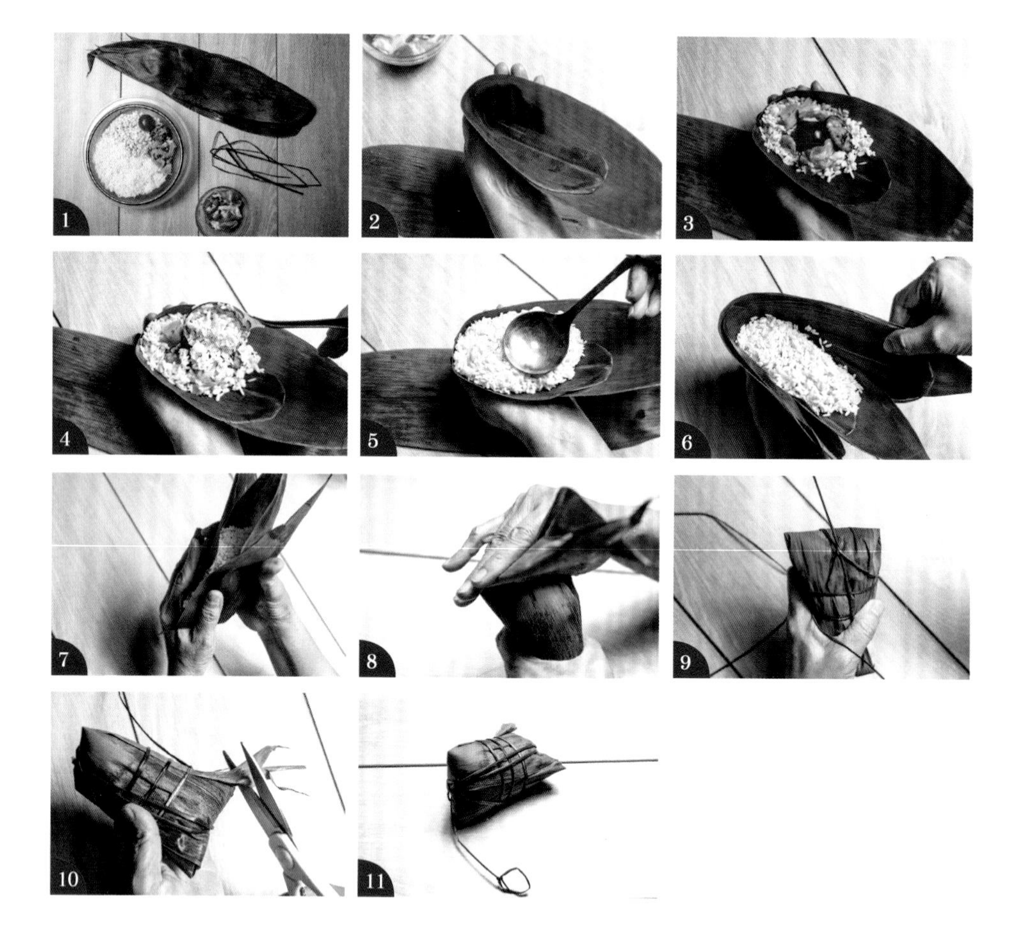

Ingredients

24 bamboo leaves
8 straws
600 g glutinous rice
1 tsp of salt
175 g green split beans
8 Chinese mushrooms
275 g pork belly
40 g dried shrimps
4 dried scallops
4 salted egg yolks

Marinade for pork belly

1 tsp salt
1/2 tsp five-spice powder
dash of sesame oil

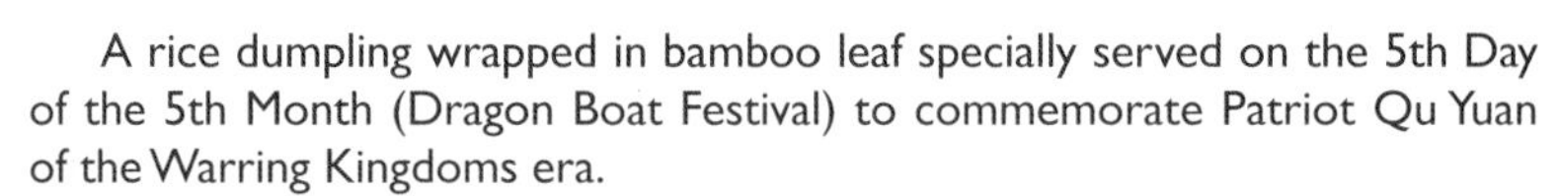

A rice dumpling wrapped in bamboo leaf specially served on the 5th Day of the 5th Month (Dragon Boat Festival) to commemorate Patriot Qu Yuan of the Warring Kingdoms era.

Method

1. Prepare the ingredients accordingly: soak and trim the bamboo leaves; soak the straws thoroughly; rinse and drain the glutinous rice and mix in 1 tsp of salt. Rinse and drain the green split beans. Picture 1
2. Hold two pieces of bamboo leaves together and fold to form a leave-shape bowl, hold in position. Picture 2
3. Spoon in a layer of glutinous rice and green split beans first, add other fillings and finish with a tbsp of glutinous rice. Pictures 3~5
4. Keep holding the dumpling with the left hand, aside in the third piece of bamboo leave in the front. Picture 6
5. Hold the dumpling upright and wrap with the left hand to seal the opening. Fold the tip part downward and hold the dumpling in position. Pictures 7~8
6. Tie and fix dumpling in shape with a long straw. Add all wrapped dumplings in a large pot of boiling water, boil for 2 hours. Drain, untie and serve. Pictures 9~10

Practical Tips

- Tie all rice dumplings in one thick straw for easy carrying or make a small loop at each end of dumpling. Picture 11

雙黃蓮蓉月
Chinese Moon Cakes

份量：5 個 / Makes 5 Cakes

皮料

麵粉 175 克
吉士粉 1 湯匙
花生油 3 湯匙
糖膠 4 湯匙
普洱濃茶 1 湯匙
鹼水 2 滴

餡料

白蓮蓉 600 克
鹹蛋黃 10 個

掃皮料

生油、蛋液（調勻）

做法

1. 將皮料混合，拌成軟粉糰，分成 5 份，搓圓，候用。
2. 白蓮蓉分成 5 份，搓圓，中間按孔，包入鹹蛋黃 2 個，埋口，搓圓。圖 1
3. 預熱焗爐至 190°C；餅模灑上一層薄粉；焗盤掃油及在案枱上灑粉。
4. 用木棍將麵皮擀成圓薄餅皮，包入蓮蓉餡一份，埋口，搓圓，放入餅模，壓平，力度要適中。圖 2
5. 敲出月餅，用噴水壺將餅略噴濕，放入焗爐先焗 10 分鐘，取出，掃上掃皮料，再焗 30 分鐘。圖 3
6. 取出月餅再掃一次皮，再焗 10-15 分鐘至餅呈金黃色為止。
7. 取出待冷透後入罐，存放 2-3 天回油後即可享用。

心得

- 新購之月餅模清洗乾淨後，先用竹籤將餅模旁之小孔通一通及用竹籤塞着小孔，注入生油浸一夜至餅模潤透為止。用吸油紙或油布印去表面油質便可使用。
- 蓮蓉及皮之大小須依月餅模之大小來確定。如製純蓮蓉月或減少鹹蛋黃，蓮蓉的份量便要增加。

Dough

175 g plain flour
1 tbsp custard powder
3 tbsp peanut oil
4 tbsp golden syrup
1 tbsp Chinese dark tea
2 drops alkali water

Filling

600 g lotus seed paste
10 salted egg yolks

To glaze

oil mix with beaten egg

A signature item for the Full Moon Festival (Mid Autumn Festival). Traditionally containing lotus seed paste and salted egg yolks. The roundness of the cakes symbolizes family union and wholeness. It also carries a historical story.

Method

1. Mix ingredients for dough, knead to a smooth dough, divide into 5 portions, shape into ball shapes.
2. Divide lotus seed paste into 5 portions, press and wrap in two salted egg yolks in each portion, seal and shape well. Picture 1
3. Preheat oven to 190°C; dust mooncake mould with flour, grease baking sheet and flour pastry board.
4. Roll each piece of dough to a thin round, wrap in a portion of filling. Seal and shape. Put in the prepared mould. Apply pressure evenly. Picture 2
5. Unmould and arrange on baking sheet, spray slightly with water. Bake for 10 minutes, take out and glaze with oil and beaten egg. Bake for another 30 minutes. Picture 3
6. Take out cakes and glaze again. Bake for 10-15 minutes until golden brown in colour.
7. Cool and keep in air-tight tins for 2-3 days. Serve when a nice glossy colour is formed.

Practical Tips

- To treat a new wooden mould, clean and block the two holes on the sides of mould with toothpicks. Fill mould with oil and soak for overnight until mould is well greased. Wipe with kitchen paper before use.
- Sizes of filling and pastry should scale to the size of mould. If less salted egg yolk is used, weight of lotus seed paste should be increased.

壽桃
Peach-shaped Buns

壽桃
Peach-shaped Buns

份量：10 個 ／ Makes 10 buns

材料

低筋麵粉 100 克
中筋麵粉 20 克
砂糖 5 克
油 1/2 茶匙
泡打粉 3 克
快速乾酵母 2 克
暖水 75 毫升
紅菜頭汁少許
蓮蓉 150 克

Ingredients

100 g low gluten flour
20 g plain flour
5 g sugar
1/2 tsp oil
3 g baking powder
2 g active dried yeast
75 ml warm water
a little pink colouring (beetroot juice)
150 g lotus seed puree

做法

1. 低筋麵粉、中筋麵粉和泡打粉一同篩勻，加入砂糖、油及快速乾酵母拌勻，注入暖水，搓勻；放案板上搓撻 3-5 分鐘，置和暖處發酵 30 分鐘。
2. 取出，用拳頭壓走氣體，輕手搓摺麵皮至軟滑不黏手，搓長分成 10 份小麵糰。圖 1~2
3. 蓮蓉分成 10 份，包入小麵糰中，搓成上尖下圓的小壽桃，靜置蒸籠內再發酵 15 分鐘。
4. 開火，凍水開始蒸壽桃 20 分鐘，取出，用刀背在壽桃尖端略按，用小牙刷趁熱噴上紅菜頭汁，即成。圖 3~5

心得

- 第一次發酵的時間約需 30 分鐘，若中途發現麵糰釋出大量氣孔或沒大反應，則酵盆／鍋內之溫度可能是過熱或過冷。熱天時，亦可放室溫（攝氏 26-28 度）自然發酵。
- 蒸小壽桃前，可用刀背預先按一小裂紋作記。圖 6

The 'Peach-shaped Buns' that look like a heavenly fruit representing longitivity, served on birthdays especially of elderlies as a toast to long life.

Method

1. Sieve low gluten flour, plain flour and baking powder together, mix in sugar, oil and yeast, add warm water to bind mixture to a soft dough, knead thoroughly in floured board for about 3-5 minutes. Leave dough in a warm, enclosed pan or deep bowl to prove for 30 minutes until well risen.
2. Take out, press dough to release the gas, knead slightly until smooth and non sticky, divide dough into 10 equal portions. Pictures 1~2
3. Divide lotus seed puree into 10 portion, wrap each portion into each piece of dough, seal to form a peach shape. Leave in steaming rack to prove for another 15 minutes.
4. Heat steamer, start timing before water is heated for 20 minutes. Take out and mark a slit with the back of a knife at the pointed side on each of the buns while still hot. Sprinkle in colouring with a tooth brush. Serve hot. Pictures 3~5

Practical Tips

- The first proof of dough need about 30 minutes. Adjust temperature of heat in the enclosed pan/ bowl if too many bubbles formed too early within time. In hot weather can just leave dough to prove at room temperature (26-28°C)
- Can premark the small slits on buns before steaming. Picture 6

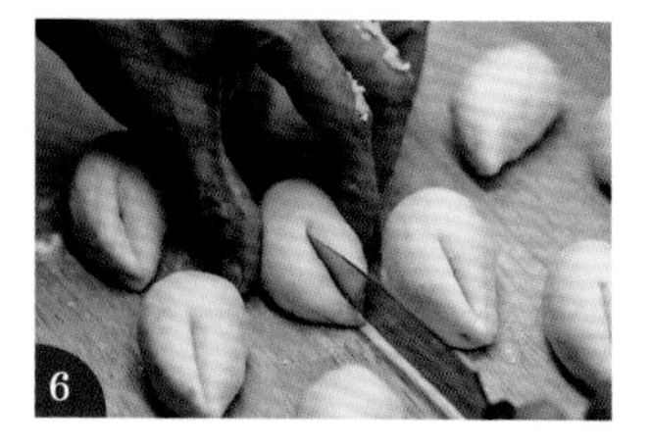

豬腳薑蛋醋
Chinese Assorted Pickles

材料

甜醋 4 樽
酸醋 1 樽
薑 900 克
豬手（斬件）900 克
雞蛋（烚熟、去殼）12 個
鹽 1/4 湯匙

Ingredients

4 bottles black sweetened vinegar
1 bottle black rice vinegar
900 g ginger
900 g pig's trotters
12 hard-boiled eggs, shelled
1/4 tbsp salt

做法

1. 用慢火將甜醋燒滾。
2. 薑去皮洗淨，吸乾水分，拍鬆，用白鑊炒乾去除多餘水分，灑入幼鹽炒片刻，盛起，加入甜醋內慢火煲 1 小時。
3. 豬手洗淨，放大滾水內滾數分鐘去除羶味，盛起過冷，用布吸乾水分。
4. 將 1 樽酸醋加入薑醋內，翻滾後加入豬手，慢火煲半小時，離火，待冷即可貯存。

心得

- 薑醋若需久存，將薑及豬手吸至極乾，否則多餘之水分會令薑醋易變壞；煲蓋之倒汗水也易使薑醋變質，所以煲薑醋之器皿宜用瓦煲。雞蛋適宜於食用 2 天前才加入薑醋內滾 10-15 分鐘後離火，享用時才加熱進食，這樣質感才適中不會過硬。

A nourishing and appetizing broth made of a juicy blend of sweet and sour vinegar consumed by mothers after 12 days of baby delivery, to help replenish energy and strength, and to celebrate the birth of a new member to the family.

Method

1. Heat sweetened vinegar slowly until boils.
2. Skin, dry and crush ginger, stir in dry heated wok to dry thoroughly, sprinkle in salt, keep stirring for a while, dish and add into sweetened vinegar, simmer for 1 hour.
3. Blanch pig's trotters in boiling water for a few minutes, rinse and towel-dry.
4. Add rice vinegar to ginger pickles, when boils, add trotters and simmer for 1/2 hour. Cool and store.

Practical Tips

- Water vapour will affect the storage life of pickles, all ingredients should be well-dried before added into vinegar. To prevent condensed water vapour, clay pots are of good choice. For best texture, add hard-boiled eggs (re-boil for 10-15 minutes) 2 days before serving.

香炸小食
Deep-fried Snacks

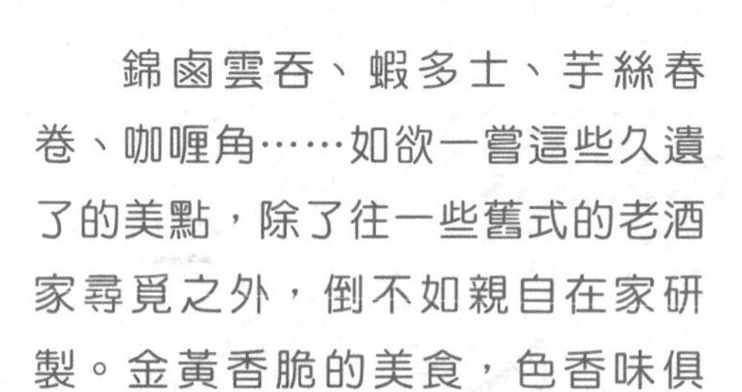

錦鹵雲吞、蝦多士、芋絲春卷、咖喱角……如欲一嘗這些久遺了的美點，除了往一些舊式的老酒家尋覓之外，倒不如親自在家研製。金黃香脆的美食，色香味俱全，往往一望已令人食指大動。

「小吃」多滋味，自家製的多吃幾件未必壞肚皮；因為如果油溫控制得宜，往往用油量未必需要太多。熟能生巧，多了解食材的耐熱及吸油性，火候拿捏得精準，可打破炸物油膩的觀念。不妨每次輕量製作，可避免忍不住口而多吃了幾件的懊悔。甚麼叫分甘同味、適可而止，由每滴垂涎開始！久不久脆一脆，更可為生活增添一點情趣。

咖喱角
Curry Beef Triangles

咖喱角
Curry Beef Triangles

份量：12 件 / Makes 12 pieces

材料

急凍春卷皮 12 塊
絞碎牛肉 100 克
洋葱半個
咖喱醬 1 滿湯匙

調味料

鹽 1/4 茶匙
糖 1/2 茶匙
生抽 1 茶匙
麻油少許

生粉水

生粉 2 茶匙
水 1 湯匙
（調勻）

黏口用

麵粉 1 湯匙
水 1 湯匙
（調勻）

做法

1. 將春卷皮剪成 4-5 厘米闊之長條，蓋好候用；牛肉調味；洋葱切碎。
2. 用 1 湯匙油起鑊，爆香咖喱醬，加入洋葱碎及牛肉，炒片刻用生粉水埋芡，盛起待涼，放入雪櫃冷藏。
3. 將春卷皮之一端摺兩下，做成一個三角袋形，放入牛肉餡，向前覆摺，最末端用麵粉糊埋口，即可放入中火油內炸脆。圖 1~3

心得

- 未完全解凍的春卷皮很易被撕爛，要待春卷皮軟化後才可逐塊分開。

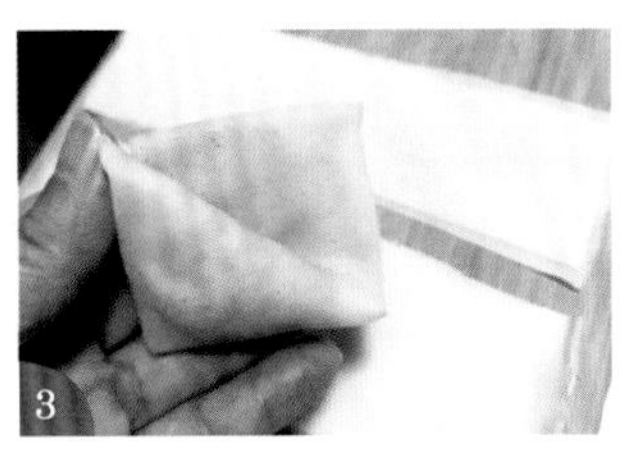

Ingredients

12 sheets frozen spring roll wrappers
100 g minced beef
1/2 onion
1 level tbsp curry paste

Seasonings

1/4 tsp salt
1/2 tsp sugar
1 tsp light soy sauce
dash of sesame oil

Cornstarch Solution

2 tsp cornstarch
1 tbsp water
} mix well

For Sealing

1 tbsp plain flour
1 tbsp water
} mix well

Method

1. Trim spring roll wrappers to long strips of 4-5 cm wide, cover and set aside. Season minced beef. Dice the onion.
2. Heat 1 tbsp oil in wok, sauté curry paste, add diced onion and seasoned minced beef, stir to cook well. Thicken with cornstarch solution. Dish, cool and chill.
3. Fold each end of stripped wrappers twice to form a triangle pocket, add 1 tsp of filling, turn and fold to form triangle puff. Seal well with the flour mixture. Deep-fry till crispy. Serve hot. Pictures 1~3

Practical Tips

- Spring roll wrappers should be well defrosted for easy separating.

材料

腐皮 1 塊
蝦膠 225 克
蛋白 1 個

調味料

胡椒粉少許
鹽 1/4 茶匙
生粉 2 茶匙

Ingredients

1 bean curd sheet
225 g shrimp paste
1 beaten egg white

Seasonings

shakes of pepper
1/4 tsp salt
2 tsp cornstarch

腐皮蝦卷
Crispy Bean Curd Rolls

份量：4 件 / Makes 4 pieces

做法

1. 腐皮用濕布抹淨，剪去硬邊，修成 4 塊正方形之腐皮，用布蓋好候用。
2. 蝦膠調味，打至起膠後置雪櫃內冷藏數小時。
3. 蝦膠分成 4 份；腐皮逐塊以菱形放平，放一份蝦膠於中央，先將兩邊摺入，再自前方捲起，以少許蛋白埋口。
4. 用中火油將腐皮卷炸脆即成，趁熱享用。

心得

- 若要蝦膠爽口彈牙，必須選購新鮮的中蝦，去殼、去腸後要用生粉略擦，再沖去潺滑之黏物及吸乾水分。
- 拍成蝦膠後，最好放雪櫃內冷藏半日或過夜。

Method

1. Wipe bean curd sheet clean, remove hard rims, trim into 4 squares. Cover with damp cloth.
2. Season shrimp paste, stir well. Chill for several hours.
3. Divide shrimp paste into 4 portions, line bean curd sheet flat, put a portion of shrimp paste in the centre. Fold from the sides and roll forward, seal with egg white.
4. Deep-fry bean curd rolls over medium heat until crispy. Serve hot.

Practical Tips

- To prepare a crunchy shrimp paste, use medium fresh prawns. Rub shelled and deveined prawns with cornstarch, rinse and dry through before crushing.
- Chill shrimp paste for 1/2 to 1 day before use.

芋絲春卷
Spring Rolls

份量：12 條 / Makes 12 rolls

材料

春卷皮 6 塊
芋頭 80 克
喼汁 1 小碟

調味料

五香粉少許
鹽 1/4 茶匙
麻油少許

麵粉糊

麵粉 1 湯匙
水 2 湯匙
（調勻）

做法

1. 春卷皮對角剪成 12 塊三角形。
2. 芋頭去皮切絲，放滾油內炸脆，撈起瀝油，攤凍，拌入調味料，分成 12 份。看附圖
3. 將三角形春卷皮鋪平，尖端向外，放上芋頭絲 1 小份，每份覆摺包成卷狀，開口處用麵粉糊黏實，置熱油內炸脆，與喼汁同上，熱食。

心得

- 春卷皮易被風乾，使用前宜用濕布或保鮮紙蓋好。

Ingredients

6 spring roll wrappers
80 g taro
a small plate of worcestershire sauce

Seasonings

a little five-spice powder
1/4 tsp salt
dash of sesame oil

Flour Paste

1 tbsp plain flour
2 tbsp water
} mix well

Method

1. Cut spring roll wrappers diagonally to get 12 triangles.
2. Peel and shred taro, deep fry in hot oil till crispy, drain, cool and sprinkle in seasonings. As shown
3. Lie the triangle wrappers flat with pointed ends facing outwards, wrap in a portion of taro fillings, fold from the two sides, roll up and seal open ends with flour paste. Deep fry spring rolls in hot oil until crispy. Dish and serve hot with worcestershire sauce.

Practical Tips

- Spring roll wrappers dry up easily. Keep covered with damp cloth or plastic wrap when not in use.

蝦多士
Shrimp Toast

份量：12 件 / Makes 12 pieces

材料

蝦膠 300 克
中蝦 6 隻
蛋白 1 湯匙
方包 2 片
白芝麻少許
喼汁 1 小碟

調味料

鹽 1/4 茶匙
胡椒粉少許
生粉 2 茶匙

Ingredients

300 g shrimp paste
6 prawns
1 tbsp egg white
2 slices white bread
a little white sesame
small plate of worcestershire sauce

Seasonings

1/4 tsp salt
shakes of pepper
2 tsp cornstarch

做法

1. 蝦膠加入調味料，順一方向拌勻至起膠，分成 12 份。
2. 中蝦去殼、去腸，洗淨，吸乾水分，橫切開二，拌入蛋白。
3. 方包切成 12 小件，每件釀上蝦膠 1 份，中間黏上中蝦肉半隻，用刀固定好，黏少許白芝麻於蝦膠上。
4. 置中火油鍋內將蝦多士炸透及呈金黃色，撈起吸乾油分，與喼汁同上，熱食。

心得

- 蝦膠在使用前，置雪櫃內冷凍 1-2 小時左右，會更彈牙、更爽口。
- 方包置雪櫃內雪至硬身或使用隔夜方包，炸後的多士會更鬆脆。

Method

1. Add seasonings to shrimp paste, mix in one direction until very sticky, divide into 12 portions.
2. Shell and devein prawns, clean and pat dry, cut into halves crosswisely, mix in egg white.
3. Cut white bread in 12 pieces, spread shrimp paste on the white bread, centre with a halved prawn, fix well in position with a knife. Sprinkle sesame seed on top of shrimp paste.
4. Deep-fry shrimp toasts in medium heat until cooked and crispy, drain. Dish and serve with worcestershire sauce.

Practical Tips

- Chill the seasoned shrimp paste for an hour or two for a crunchy effect.
- Use stale bread for fried toast to enhance crispiness.

錦鹵雲吞
Jumbo Wontons with Sweet and Sour Sauce

份量：15 件 / Makes 15 pieces

材料

錦鹵雲吞皮 15 塊
蝦肉 225 克

調味料

鹽 1/8 茶匙
生粉 1 茶匙
胡椒粉少許
蛋黃 1/2 個

錦鹵料

蝦肉 100 克
叉燒 50 克
鮮魷魚 1 隻
洋葱 1/2 個
青椒 1/2 個
尖紅椒 1 隻
菠蘿 2 片

甜酸芡

水 250 毫升
白醋 250 毫升
茄汁 125 毫升
喼汁 1 茶匙
糖 1/3 杯
鹽 1/2 茶匙

生粉水

生粉 1.5 湯匙
水 2 湯匙
〉調勻

做法

1. 將 225 克蝦肉吸乾水分後切粒，加入調味料拌勻並冷凍片刻，每塊雲吞皮包入少量餡料圖 1，於餡邊緣沾水少許埋口圖 2，對摺成多角形圖 3~5，置熱油內炸至鬆起及脹大，炸脆盛起。圖 6
2. 預備錦鹵料：將蝦肉出水；叉燒切片；魷魚洗淨，切花後切件，出水；洋葱、青椒、尖紅椒及菠蘿切塊；甜酸芡拌勻待用。
3. 燒油 2 湯匙，先爆香洋葱件、紅椒及青椒，炒片刻，加入其他材料，炒勻，倒入甜酸芡，煮滾用生粉水埋芡，煮稠，上碟。
4. 將炸好之雲吞蘸上甜酸芡，趁熱供吃。

心得

- 錦鹵雲吞皮為特製之較大塊雲吞皮，內加適量食用臭粉，只宜油炸用，食其鬆脆之風味及蘸以特製之錦鹵芡，美味無窮。圖 7
- 一般製麵店只供酒樓食館訂製錦鹵雲吞皮，家庭少量製作，可到粉麵店預先訂購，一般 1-2 天可有交易。

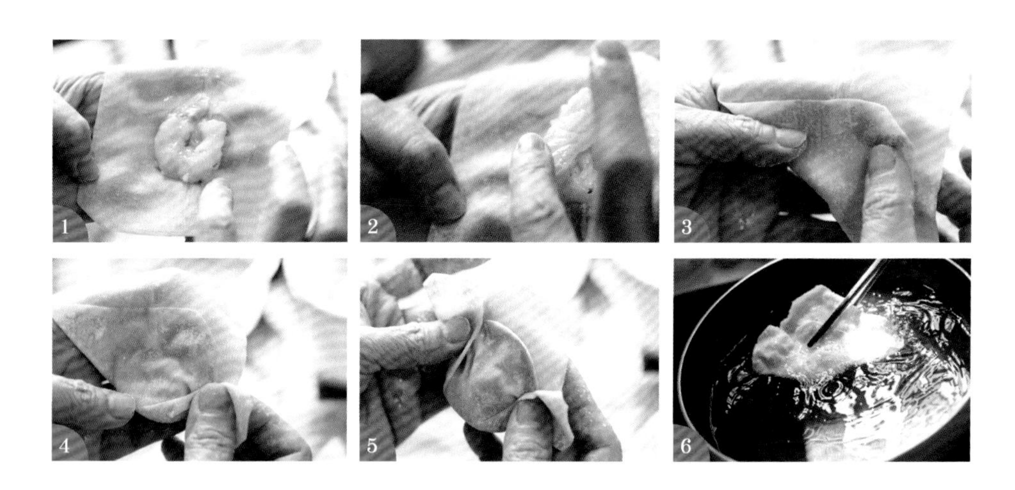

Ingredients

15 sheets wonton wrappers
225 g shelled prawns

Seasonings

1/8 tsp salt
1 tsp cornstarch
pinch of pepper
1/2 egg yolk

Ingredients for Sweet and Sour Sauce

100 g shelled shrimps
50 g roast pork
1 squid
1/2 onion
1/2 green pepper
1 red chilli
2 slices pineapple

Sweet and Sour Sauce

250 ml water
250 ml white vinegar
125 ml tomato ketchup
1 tsp worchestershire sauce
1/3 cup sugar
1/2 tsp salt

Thickening

1.5 tbsp cornstarch
2 tbsp water
} mix well

Method

1. Dry and dice 225 g shelled prawns, add seasonings, mix and chill. Wrap a little filling in each wonton wrapper Picture 1, seal edge with water Picture 2 and fold diagonally Pictures 3~5. Deep-fry wontons in hot oil until golden brown and crispy. Remove and drain. Picture 6
2. Blanch 100 g shelled shrimps; slice the roast pork; score and cut the squid into pieces and blanch. Cut onion, green pepper, red chilli and pineapple slices into pieces. Mix sweet and sour sauce well.
3. Heat 2 tbsp of oil in wok, sauté onion, green pepper and red chilli, stir well, add other ingredients and sweet and sour sauce, when boils, thicken with cornstarch solution.
4. Dish and serve hot with fried Jumbo Wontons.

Practical Tips

- Wonton wrappers for deep-fry are specially made for a short and crispy result because of the edible ammonia powder added in the flour dough. The Jumbo Wontons are specially served with sweet and sour sauce. Picture 7
- Wrappers for Jumbo Wontons should be ordered 1-2 days in advance from special noodles manufacturers.

7

蜆蚧鯪魚球
Golden Fish Balls

份量：9 個 / Makes 9 pieces

材料

絞碎鯪魚肉 275 克
果皮 1 小塊
鹽 1/2 茶匙
粟粉 1 湯匙
生抽 1/2 茶匙
糖 1/2 茶匙
胡椒粉少許
蛋白 1/2 個
水 2 茶匙

蘸料

蜆蚧醬

做法

1. 果皮浸軟，去瓤後切碎。
2. 將所有材料拌勻，打至起膠，置雪櫃內冷藏 2-4 小時。
3. 魚膠分成 9 份，搓成丸狀，輕滾上生粉一層，置熱油內炸至魚球熟及呈金黃色。
4. 撈起魚球，與蜆蚧醬同上，熱食。

心得

- 蜆蚧醬可到舊式之雜貨店、醬料專門店及中國國貨公司購買。

Ingredients

275 g minced dace
1 small piece dried tangerine peel
1/2 tsp salt
1 tbsp cornstarch
1/2 tsp light soy sauce
1/2 tsp sugar
shakes of pepper
1/2 beaten egg white
2 tsp water

To serve

preserved clam meat paste

Method

1. Soak the dried tangerine peel till soft. Remove the pith and chop finely.
2. Mix all ingredients together. Stir until firm and sticky. Chill in the refrigerator for 2-4 hours.
3. Divide fish paste into 9 portions. Shape into round balls, slightly coat with cornstarch, deep-fry in hot oil until cooked and golden brown.
4. Dish and serve fish balls with preserved clam meat paste.

Practical Tips

- Preserved clam meat paste is available in Chinese Emporiums and some of the traditional grocers.

椒鹽豆腐
Spicy Bean Curd

份量：16 件 / Makes 16 pieces

材料

硬豆腐 2 件
淮鹽 1 茶匙

脆漿料

自發粉 100 克
冰水 225 毫升
葱粒 1 湯匙
紅椒碎 1 湯匙
鹽 1/2 茶匙

蘸料

淮鹽少許

做法

1. 硬豆腐一開八，共十六件，灑上適量淮鹽略醃。
2. 將脆漿料調勻，靜置片刻。
3. 豆腐吸乾水分，黏上脆漿料，置熱油內炸至鬆脆，上碟，蘸餘下的淮鹽享用。

心得

- 此小食宜即炸即食，因豆腐的水分及熱氣會將脆皮軟化。但如果用大火將麵漿炸透至挺硬，外皮的鬆脆度也能保持一段時間。

Ingredients

2 pieces firm bean curd
1 tsp spicy salt

Batter

100 g self-raising flour
225 ml iced water
1 tbsp diced spring onion
1 tbsp chopped red chilli
1/2 tsp salt

To Dip

a little spicy salt

Method

1. Quarter firm bean curds, cut into half to get 16 pieces. Season well with a little spicy salt.
2. Mix ingredients for batter, set aside for a while.
3. Drain bean curd, coat with batter, deep-fry in hot oil until golden brown. Drain and serve with remaining spicy salt.

Practical Tips

- The crispy coating tends to become limp in a short time because of the bean curd's softness and hot steam enclosed. Deep-fry the bean curd over high heat until a hard crust is formed helps to keep the crunchiness of the coating.

炸釀豆腐泡
Crispy Stuffed Bean Curd Puff

份量：30 件 / Makes 30 pieces

材料

鯪魚肉 350 克
蝦膠 100 克
豆腐泡 15 個
喼汁適量（伴食用）

調味料

鹽 3/4 茶匙
糖 1/4 茶匙
粟粉 2 茶匙
麻油少許
胡椒粉少許

做法

1. 鯪魚肉與蝦膠拌勻，加調味料，撻至起膠。
2. 將豆腐泡一開二，反轉裏面向外成袋形，釀入適量餡料，放熱油中炸至香脆，撈起，用喼汁伴食。

心得

- 蘸料可以用茄汁或沙律醬代替。

Ingredients

350 g minced fish
100 g shelled prawns (mashed)
15 bean curd puffs
a little worchestershire sauce (to dip)

Seasonings

3/4 tsp salt
1/4 tsp sugar
2 tsp cornstarch
dash of sesame oil
shakes of pepper

Method

1. Mix minced fish and mashed prawns together, season, stir to a sticky paste.
2. Cut bean curd puffs into halves, turn inside out to form a pocket, stuff with filling, shape and deep-fry in hot oil until cooked and crispy. Drain and serve with worchestershire sauce.

Practical Tips

- Tomato ketchup or salad dressing can be used as a dip instead of worchestershire sauce.

炸饅頭
Crispy Sweet Buns

份量：10 個 / Makes 10 buns

材料

麵粉 225 克
砂糖 1 湯匙
豬油 1 湯匙
乾酵母 1/2 茶匙
泡打粉 1/2 茶匙
暖水 125 毫升

蘸料

煉奶少許

做法

1. 乾酵母及泡打粉加入暖水內和勻候用。
2. 麵粉與砂糖同篩入大碗內，中間開穴，倒入酵母水及豬油，拌勻搓至軟滑，蓋好靜置 1 小時。
3. 將麵糰搓成長條，用刀切成小饅頭，再蓋好待 10 分鐘，使其質地鬆軟，置蒸籠內大火蒸 15 分鐘。
4. 取出饅頭，待冷卻後置中火油內炸至呈金黃色即可，煉奶伴食。

心得

- 麵糰待發的時間不定，須視乎室溫而定，如見輕微發起及軟身即可。如見麵糰有小氣泡狀，表示發酵時間過長，蒸出來之饅頭會不夠理想。

Ingredients

225 g plain flour
1 tbsp castor sugar
1 tbsp lard
1/2 tsp dry yeast
1/2 tsp baking powder
125 ml warm water

To Dip

a little condensed milk

Method

1. Mix dry yeast and baking powder in warm water.
2. Sieve plain flour and castor sugar in a deep bowl, make a well in the centre, add yeast solution and lard, mix and knead to form a soft dough. Cover and proof for 1 hour.
3. Knead and roll dough to a long sausage form, cut into mini buns. Keep buns in a bamboo steamer, cover for 10 minutes for softening the texture. Steam over strong heat for 15 minutes.
4. Cool and deep-fry buns until crispy and golden brown. Serve with condensed milk.

Practical Tips

- When tiny air bubbles are found in dough, it is over risen. Dough is ready when it becomes springy and soft. Time for proving depends upon room temperature.

沙律海鮮卷
Fried Fruity Seafood Parcels

份量：8 件 / Makes 8 parcels

材料	Ingredients
威化紙 8 張	8 sheets rice paper
中蝦 4 隻	4 prawns
帶子 4 粒	4 scallops
蟹柳 4 條	4 crab meat sticks
草莓 6 顆	6 strawberries
沙律醬 3 湯匙	3 tbsp salad dressing
蛋白 2 個（拌勻）	2 egg white, whisked
日式麵包糠 1 量杯	1 cup white breadcrumbs

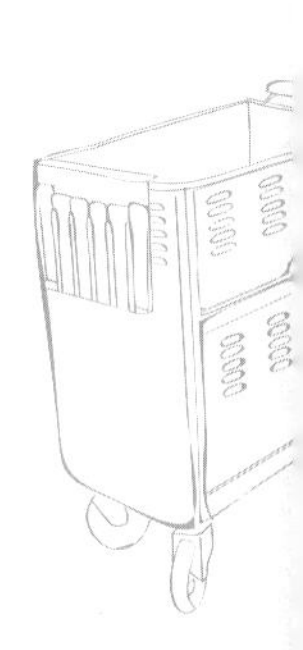

做法

1. 中蝦去殼、去腸，洗淨抹乾水分，橫切開二，出水，瀝乾。
2. 帶子洗淨吸乾水分，橫切開二，出水，瀝乾。
3. 蟹柳切段。
4. 草莓洗淨切粒，加入沙律醬及少許鹽，拌勻。
5. 砧板抹乾，威化紙對摺重疊至中間位置，放入蟹柳、中蝦及帶子肉各一圖 1~2，加 1 茶匙草莓沙律圖 3，快手包捲成長方形海鮮卷圖 4~8，用手逐件放入蛋白液，立刻撈起沾上麵包糠。圖 9~11
6. 用中火燒熱油，放入海鮮卷炸至金黃色，撈起，瀝乾油分，上碟，熱食。

心得

- 威化紙沾水即溶，砧板及手要徹底抹乾才開始捲入海鮮，製作時動作要迅速及連貫。
- 可用少許麵包糠測試油溫：如麵包糠向下沉，即表示油溫未夠熱；如立刻浮起，即可加入海鮮卷。

Method

1. Shell and devein prawns, pat dry, cut into 2 pieces lengthwisely, blanch and drain.
2. Clean and dry scallop, cut into 2 pieces lengthwisely, blanch and drain.
3. Cut crabmeat sticks into segments.
4. Clean and dice strawberries, add salad dressing and a pinch of salt, mix well.
5. Wipe chopping board very dry, double fold rice paper to centre line, wrap in assorted seafood Pictures 1~2 , top with a teaspoonful of strawberry salad. Picture 3 Fold and wrap into shape of a parcel. Pictures 4~8 Quickly dip in beaten egg white followed by even coating of white breadcrumbs. Pictures 9~11
6. Heat oil, deep-fry seafood parcels over medium heat till golden brown and crispy, drain off excessive oil, dish and serve hot.

Practical Tips

- Keep chopping board and hands very dry, actions should be quick and continuous in wrapping, rice paper melts easily on a wet surface.
- Test the readiness of oil with a little breadcrumbs which should floats quickly in the heated oil.

炸鮮奶
Fried Milk Custard

炸鮮奶
Fried Milk Custard

份量：10-12 件 / Make 10-12 pieces

材料

粟粉 40 克
鮮奶 375 毫升
蛋白 2 個（拂勻）
油 1 湯匙
鹽少許
砂糖 1/2 茶匙

脆漿

麵粉 75 克
粟粉 40 克
發粉 1 茶匙
鹽少許
油 1 湯匙
水 125 毫升

伴食

砂糖 1/2 杯

做法

1. 將粟粉及鮮奶拌勻，加油、鹽和砂糖作調味，慢火煮成糊狀，須不停攪拌，待滾。
2. 離火，分 3 次快手攪入蛋白，拌勻，倒入已塗油的方形糕盤內，待冷，置雪櫃凍藏 1-2 小時。
3. 將脆漿料拌勻，靜置片刻。
4. 倒出鮮奶糕，切件，沾上脆漿置八成熱油內炸至香脆及呈金黃色。瀝乾油分，沾上砂糖熱食。

心得

- 此脆漿可作多種用途，如炸豬扒、雞翼等。

Ingredients

40 g cornstarch
375 ml fresh milk
2 egg whites, beaten
1 tbsp oil
pinch of salt
1/2 tsp sugar

Batter

75 g plain flour
40 g cornstarch
1 tsp baking powder
pinch of salt
1 tbsp oil
125 ml water

To serve

1/2 cup castor sugar

Method

1. Mix cornstarch with fresh milk, add oil, season with salt and sugar. Stir constantly over gentle heat until cooked.
2. Remove from heat, add beaten egg white gradually, stir between each addition. Mix well to form a smooth mixture, pour into a greased square tin. Cool and set in the refrigerator for 1-2 hours.
3. Mix batter ingredients to a thick consistency. Set aside.
4. Unmould milk custard, cut into cubes, coat with batter. Deep-fry over medium-high heat till set, crispy and golden brown. Drain, dip with castor sugar. Serve hot.

Practical Tips

- This coating batter can be used for fried pork chop and chicken wings.

精美鹹點
Savoury Snacks

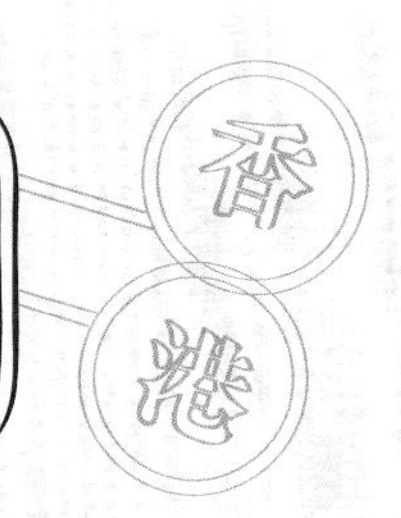

看見糯米雞、四寶雞紮、鮮竹卷、煎釀三寶、潮州粉粿等鹹點，彷彿置身於昔日的酒樓茶居，與家人嘆着一盅兩件、聽着那一邊推着點心車的姨姨的叫賣聲，與今天拿着點心紙「落單」的情況，已不再一樣。

豬皮魷魚蘿蔔、煎釀三寶是昔日街頭車仔麵檔的招牌小吃，那「掃街」文化，畢竟也養活過不少一家七、八口子的小康家庭。然而，都市的進步，飲食習慣及生活的漸趨文明，也動搖不了這系列中樸實無華、可以裹腹的非一般小吃。

糯米雞
Steamed Glutinous Rice with Assorted Meat in Lotus Leaf

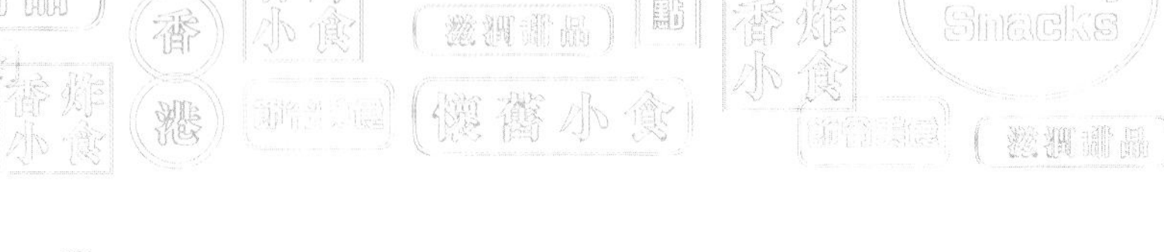

糯米雞

Steamed Glutinous Rice with Assorted Meat in Lotus Leaf

份量：2 隻 / Makes 2 pieces

材料

糯米 250 克
乾荷葉 2 塊

餡料

雞肉 80 克（切粒）
叉燒 80 克（切片）
冬菇 2 朵（浸軟）
蝦米 30 克（浸軟）
乾葱 1 粒
熟蛋（烚）1/2 個

調味料（糯米用）

鹽 1/2 茶匙
糖 1/2 茶匙
油 1 湯匙
老抽 1/2 茶匙

芡汁料

水 50 毫升
粟粉 1/2 湯匙
糖 1/2 茶匙
生抽 1/2 茶匙
老抽 1/2 茶匙
麻油 1/2 茶匙
蠔油 1 湯匙
胡椒粉少許

做法

1. 荷葉洗淨，用熱水浸至軟身，瀝乾水分。
2. 糯米洗淨，放入滾水內煮 2-3 分鐘，用冷水沖洗，瀝乾，加入調味料拌勻，再蒸約 20-25 分鐘，分成 4 份。
3. 用少許粟粉及鹽略醃雞肉粒；乾葱去皮，拍扁，切茸；熟蛋半個一開二。預備芡汁。
4. 燒熱 1-2 湯匙油，爆香乾葱茸，加入雞肉粒炒至熟透，加叉燒片、冬菇及蝦米，灒酒 1 茶匙，加入芡汁炒勻，盛起分成兩份。
5. 荷葉放平，滑面向上，掃油，加入 1 份糯米飯，放 1 份餡料於上，再放另 1 份糯米飯，覆摺成糯米雞，大火蒸 10 分鐘，趁熱進食。

心得

- 熟蛋於包糯米飯時才放入，勿與其他餡料同炒。

Ingredients

250 g glutinous rice
2 lotus leaves

Filling

80 g chicken meat, diced
80 g roast pork, sliced
2 Chinese mushrooms, soaked
30 g dried shrimps, soaked
1 clove of shallot
1/2 hard-boiled egg

Seasonings (for glutinous rice)

1/2 tsp salt
1/2 tsp sugar
1 tbsp oil
1/2 tsp dark soy sauce

Sauce

50 ml water
1/2 tbsp cornstarch
1/2 tsp sugar
1/2 tsp soy sauce
1/2 tsp dark soy sauce
1/2 tsp sesame oil
1 tbsp oyster sauce
shakes of pepper

Method

1. Rinse lotus leaves, soak in hot water until quite soft. Drain.
2. Wash glutinous rice, cook in boiling water for 2-3 minutes, rinse with cold water, drain well, add seasonings and steam for 20-25 minutes until cooked. Divide into 4 portions.
3. Marinate chicken dices with a little cornstarch and salt; crush and chop shallot; cut hard-boiled egg into 2; prepare sauce ingredients.
4. Heat oil, sauté chopped shallot, add chicken, stir fry until cooked, add roast pork, Chinese mushroom and dried shrimps, sizzle in 1 tsp wine, add sauce ingredients, stir well, dish and divide into 2 portions.
5. Lie lotus leaves flat with silvery side facing upward, grease, add 1 portion of cooked glutinous rice in the centre, then add 1 portion of assorted meat on top, sandwich with another portion of glutinous rice, fold and wrap up to form a parcel shape. Steam over high heat for 10 minutes. Serve hot.

Practical Tips

- Add egg wedges at step 5, do not stir fry with the other ingredients in pan.

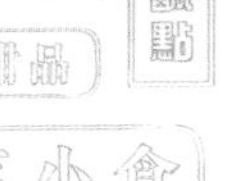

滷水牛腱

Soy Beef

材料

牛腱 450 克
薑 1 厚片
八角 1 粒
果皮 1 片
鹽 1/4 茶匙
冰糖 75 克
老抽 125 毫升
生抽 60 毫升
玫瑰露酒 1 湯匙
水 1 公升

Ingredients

450 g shin of beef
1 thick slice ginger
1 star aniseed
1 piece dried tangerine peel
1/4 tsp salt
75 g rock sugar
125 ml dark soy sauce
60 ml light soy sauce
1 tbsp Chinese rose wine
1 litre water

做法

1. 牛腱洗淨，出水後瀝乾；薑片拍鬆；果皮浸軟後去瓤。
2. 將水燒滾，加入所有材料，翻滾後收慢火燜 1.5 小時。
3. 取出牛腱，稍凍後切片上碟。

心得

- 可用少許汁料及生粉水埋芡，煮滾淋於牛腱上，食味更佳。

Method

1. Clean, blanch, rinse and drain shin of beef. Crush the ginger slice. Soak and scrape dried tangerine peel.
2. Boil water, add all ingredients, lower the heat when re-boils, simmer for 1.5 hours.
3. Take out the beef. Cool, slice and serve.

Practical Tips

- For a juicier effect, thicken the remaining gravy with a little cornstarch solution and pour on top of the sliced beef.

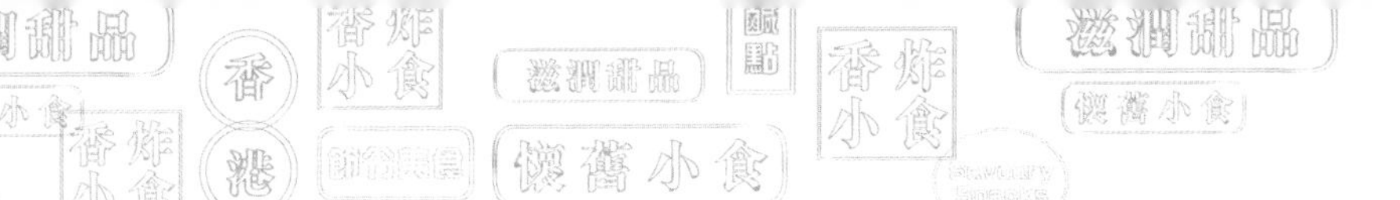

鹽水鳳爪

Stewed Spicy Chicken Claws

材料

急凍鳳爪 12 隻
紹酒 1 湯匙
麻油 1 湯匙

白滷水料

八角 1 粒
果皮 1 小片
薑 1 片
粗鹽 1 湯匙
水 1 公升

做法

1. 急凍鳳爪洗淨，剪去腳甲，置大滾水內，出水片刻，加入紹酒，大滾後過冷水，瀝乾候用。
2. 將白滷水料燒滾，加入鳳爪，慢火煮 20-25 分鐘，離火，撈起置冰水內浸泡 1-2 小時，再倒回已冷卻的白滷水中，置雪櫃內一夜。
3. 撈起鳳爪，掃上麻油，即成。

心得

- 急凍鳳爪出水時，倒入紹酒，有助去掉雪味。
- 用小火來滷鳳爪，可保持鳳爪賣相完整及更加入味。

Ingredients

12 frozen chicken claws
1 tbsp Shaoxing wine
1 tbsp sesame oil

Spicy mix

1 star aniseed seed
1 small piece dried tangerine peel
1 slice ginger
1 tbsp coarse salt
1 litre water

Method

1. Clean and blanch frozen chicken claws in boiling water with Shaoxing wine. Rinse in cold water, drain well.
2. Bring spicy mix to the boil, add chicken claws and simmer for 20-25 minutes until cooked. Drain into a bowl of ice water, soak for 1-2 hours. Drain and put back into cooled spicy mix. Chill overnight.
3. Drain, add sesame oil, mix well, serve cold.

Practical Tips

- Shaoxing wine helps to remove the icy and raw smell of chicken claws.
- Simmering in low heat helps to keep the chicken claws in good shapes and the spicy flavour to penetrate.

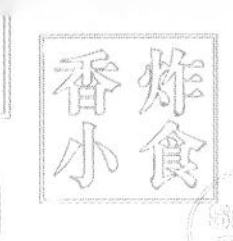

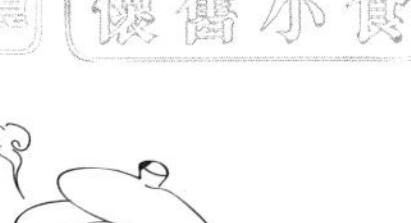

茶葉蛋
Tea Flavoured Eggs

份量：12 個 / Makes 12 eggs

材料

雞蛋 12 個
水 3 杯
八角 4 粒
桂皮 1 小片
茶葉（鐵觀音）3 湯匙
片糖 250 克
老抽 125 毫升
生抽 60 毫升
鹽 2 茶匙

Ingredients

12 eggs
3 cups water
4 star aniseed
1 small piece cassia bark
3 tbsp tea leaves (Tiguanyin)
250 g slab sugar
125 ml dark soy sauce
60 ml light soy sauce
2 tsp salt

做法

1. 將雞蛋烚熟（約 10 分鐘），敲裂蛋殼待用。
2. 將其餘材料煲滾，放入雞蛋，翻滾後收慢火煮 15 分鐘，熄火焗半小時。
3. 再將茶葉蛋煲滾，離火焗過夜。
4. 加熱後即可供吃。

心得

- 茶葉汁料需以浸過蛋面為合。

Method:

1. Hard-boil eggs for 10 minutes, crack shells.
2. Bring the rest of ingredients to boil. Add eggs, simmer for 15 minutes. Remove from heat, cover for 1/2 hour.
3. Re-boil tea-flavoured eggs and cover overnight.
4. Reheat and serve.

Practical Tips

- All cracked eggs should be well covered by the tea-leaf sauce.

潮州粉粿
Chui Chow Dumplings

份量：12 個 / Makes 12 dumplings

皮料

澄麵粉 175 克
生粉 75 克
鹽少許
滾水 375 毫升
生粉 75 克（後下）

調味料

鹽、糖各 1/4 茶匙
粟粉 1 茶匙
水 1 湯匙
油 1/2 湯匙
胡椒粉、麻油各少許

餡料

瘦肉 175 克
花生 75 克
沙葛 75 克
韭菜 50 克
菜脯 50 克

芡汁

粟粉 1 茶匙
鹽、生抽各 1/2 茶匙
水 60 毫升
五香粉少許（後下）

做法

1. 澄麵粉、生粉及鹽混合，沖入滾燙熱水，拌勻，加蓋焗片刻，再撥入75克生粉，搓至不黏手為合。
2. 瘦肉切粒後加入調味醃 20 分鐘；花生連衣炸熟；沙葛、韭菜及菜脯洗淨後分別切粒。
3. 用 1 湯匙油起鑊，加入肉粒炒熟，再加入其他餡料並埋芡，離火，灑入五香粉，取出待涼。
4. 將粉糰分成 12 等份，每份搓圓按扁，包入適量餡料，對摺成半月形，再打摺捏成粉粿形，大火蒸 20 分鐘。圖 1~2
5. 趁熱掃上少許熟油即可。

心得

- 如喜歡較煙韌之粉粿皮，可往專門售東南亞用料之店舖購買泰國生粉，也稱木薯粉，但一般家庭打芡用之生粉亦可。

1

2

Dough

175 g ungluten flour
75 g tapioca starch
pinch of salt
375 ml boiling water
75 g tapioca starch, reserved

Filling

175 g lean pork
75 g peanuts
75 g yam bean
50 g green chives
50 g salted turnip

Seasonings

1/4 tsp salt
1/4 tsp sugar
1 tsp cornstarch
1 tbsp water
1/2 tbsp oil
pinch of pepper
dash of sesame oil

Sauce

1 tsp cornstarch
1/2 tsp salt
1/2 tsp light soy sauce
60 ml water
a little five-spice powder, reserved

Method

1. Sieve ungluten flour, tapioca starch and salt together. Pour in boiling water, mix well, cover for a while. Add reserved tapioca starch, mix to form a smooth dough.
2. Dice and season the pork for 20 minutes. Deep-fry peanuts with skin on. Wash and dice yam bean, green chives and salted turnip separately.
3. Cook pork in 1 tbsp oil, add other ingredients, bind with sauce ingredients, remove from heat and sprinkle in five-spice powder, dish and cool.
4. Divide dough into 12 equal portions. Knead and press each to a round, wrap in sufficient filling, fold to form a half moon shape, shape to pleated Chiu Chow dumplings. Steam for 20 minutes over high heat. Pictures 1~2
5. Brush with cooked oil, serve hot.

Practical Tips

- The tapioca starch specially for this transparent dough is available from shops that sell south-east Asian ingredients.

四寶雞紮
Four Treasures Bean Curd Rolls

四寶雞紮
Four Treasures Bean Curd Rolls

份量：8 件 / Makes 8 rolls

材料

腐皮 1/2 塊
雞髀肉 1 件
火腿 1 厚片
浸發魚肚 1 塊
芋頭 80 克

調味料（雞肉）

鹽、糖、胡椒粉、麻油、油、生粉各少許

鹽水（芋頭）

熱水 200 毫升
鹽 1/2 茶匙

生粉水（魚肚）

生粉 1.5 湯匙
水 100 毫升
鹽 1/2 茶匙

做法

1. 腐皮用濕布抹淨，剪成 5×12 厘米之長方塊 8 片（餘下腐皮邊可留作其他用途），逐片腐皮放入熱油內炸鬆；將已炸鬆的腐皮放入水內浸軟，撈起瀝乾備用。
2. 雞髀肉切成長條，加調味醃 3 小時，蒸熟。
3. 火腿切成條狀。
4. 魚肚切塊，用生粉及鹽水浸煮片刻，使之入味。
5. 芋頭切條，用油炸熟，瀝乾油分，浸鹽水。
6. 將浸軟的腐皮攤平，包入熟雞肉、火腿條、魚肚及芋頭各一件，捲好，放碟上；用中火蒸 5-8 分鐘。上碟，熱食。

心得

- 雞紮可預先製作，置雪櫃內，隨吃隨蒸。
- 可隨意用芹菜梗紮結裝飾。

Ingredients

1/2 piece bean curd sheet
1 piece chicken thigh meat
1 thick slice cooked ham
1 piece blanched fish bladder
80 g taro

Seasonings (chicken)

each of a little salt, sugar, pepper, sesame oil and cornstarch

Salt Solution (taro)

200 ml hot water
1/2 tsp salt

Cornstarch Solution (fish bladder)

1.5 tbsp cornstarch
100 ml water
1/2 tsp salt

Method

1. Trim bean curd sheet into 8 strips of about 5x12 cm in size, deep-fry piece by piece in hot oil, lift and soak in water to soften, drain well for later use.
2. Cut chicken meat into thick strips, marinate for 3 hours. Steam to cook well.
3. Cut cooked ham into strips.
4. Cut blanched fish bladder into pieces, cook in cornstarch solution for 1-2 minutes.
5. Cut taro into thick strips, deep-fry till cooked, drain and soak in salt solution. Drain before use.
6. Lie bean curd sheet strips flat, roll in a piece of chicken, ham, fish bladder and taro to make eight Bean Curd Rolls. Steam over medium heat for 5-8 minutes. Dish and serve hot.

Practical Tips

- Prepare Four Treasures Bean Curd Rolls ahead of time, keep in refrigerator and steam before serving.
- Garnish with Chinese celery knots as desired.

鮮竹卷
Bean Curd Rolls in Oyster Sauce

份量：12 件 / Makes 12 pieces

材料

鮮腐皮 1 張
絞碎梅頭豬肉 275 克
蝦肉 75 克

調味料

鹽 1/2 茶匙
粟粉 3 茶匙
生抽 1/4 茶匙
胡椒粉少許
麻油少許
水 2 湯匙

芡汁

水 3/4 杯
蠔油 1 湯匙
老抽 1/4 茶匙
糖、胡椒粉、麻油各少許

生粉水

生粉 1 茶匙
水 1 湯匙
（調勻）

做法

1. 腐皮用濕布抹淨，去硬邊，修剪成 12 塊正方形，蓋好候用。蝦肉洗淨後切粒。
2. 將調味料拌勻，加入豬肉及蝦粒，拌至起膠，放入雪櫃冷藏一會，分成 12 份。
3. 腐皮放平，放入餡料，包成卷狀，用少許餡料或麵粉糊封口，用中大火油炸至鮮竹卷鬆起，即可撈起。見附圖
4. 將芡汁煮滾，收慢火放入鮮竹卷炆 15 分鐘，埋生粉水，收汁後即可享用。

心得

- 亦可將炸好之鮮竹卷蒸至熟透及至軟身，將芡汁煮滾埋薄芡，淋在鮮竹卷上。

Ingredients

1 bean curd sheet
275 g minced pork shoulder
75 g shelled prawns

Seasonings

1/2 tsp salt
3 tsp cornstarch
1/4 tsp light soy sauce
shakes of pepper
dash of sesame oil
2 tbsp water

Sauce

3/4 cup water
1 tbsp oyster sauce
1/4 tsp dark soy sauce
pinch of sugar
shakes of pepper
dash of sesame oil

Thickening

1 tsp cornstarch
1 tbsp water
(mix well)

Method

1. Wipe and trim bean curd sheet into 12 squares. Cover. Wash and dice the shelled prawns.
2. Season pork and diced prawns, stir well as a paste. Chill for a while. Divide into 12 portions.
3. Add a portion of filling on each bean curd sheet, wrap and seal with a little flour paste or meat mixture. Deep-fry over medium-high heat until bean curd rolls puffs immediately. Drain. As shown
4. Heat sauce ingredients, add bean curd rolls and stew over low heat for 15 minutes. Thicken, stir well.

Practical Tips

- An alternate method is to steam the crispy bean curd rolls till thoroughly cooked and become soft. Cook the sauce, thicken and pour on top of the bean curd rolls.

薯仔餅
Potato Cakes

份量：12 件 / Makes 12 pieces

材料

薯仔 450 克
冬菇 4 朵
蝦米 2 湯匙
臘腸 1/2 條
葱粒 1 湯匙
雞蛋 1 個（拂勻）
粟粉 3 湯匙

調味料

鹽 1/2 茶匙
粟粉 1 湯匙
胡椒粉、
麻油各少許

做法

1. 薯仔洗淨，連皮放入過面滾水內焓約 20 分鐘至熟，取出去皮，趁熱壓成薯茸。
2. 冬菇及蝦米浸軟，切細粒；臘腸蒸熟切幼粒，與葱粒一同放入薯茸中，調味後分成 12 等份。
3. 將每份薯茸料搓圓按扁，捏成三角形或圓餅形，掃上蛋液，輕拍上粟粉，置中火油鑊煎至薯餅兩面呈金黃色即可供吃。

心得

- 可將一半薯仔壓茸，一半切成粗粒互相混合，一新口感。

Ingredients

450 g potato
4 dried Chinese mushrooms
2 tbsp dried shrimps
1/2 Chinese sausage
1 tbsp diced spring onion
1 egg, beaten
3 tbsp cornstarch

Seasonings

1/2 tsp salt
1 tbsp cornstarch
pinch of pepper
dash of sesame oil

Method

1. Scrub potato, put in a large pot of boiling water, boil for 20 minutes approximately or till cooked, skin and mash to form potato purée.
2. Soak and dice Chinese mushrooms and dried shrimps; steam Chinese sausage until cooked, dice finely and add into potato purée with diced spring onion or other ingredients, season, mix and divide into 12 equal portions.
3. Shape each portion of potato mixture into a round or triangular patty, press slightly, coat with beaten egg and cornstarch, shallow-fry over medium heat until golden brown on both sides. Serve.

Practical Tips

- Mash half of the cooked potato and dice the rest for a variety of texture.

蓮藕餅
Lotus Root Patties

份量：12 件 / Makes 12 pieces

材料

蓮藕 225 克
絞碎鯪魚肉 100 克
冬菇 2 朵
蝦米 2 湯匙
葱粒 1 湯匙
蛋黃 1 個
麵粉 1 湯匙
鹽 1/4 茶匙

調味料

鹽 1/2 茶匙
糖 1/2 茶匙
麻油、胡椒粉各少許

做法

1. 蓮藕去皮刨成長絲；冬菇及蝦米分別浸軟後切粒，用少許油爆香盛起。
2. 魚肉與調味料拌勻，加入其餘材料拌勻，分成 12 等份。
3. 平底鑊燒熱，加油 2 湯匙，將蓮藕料逐份放入，用鑊鏟按平使成圓餅形，用中大火煎至兩面皆呈金黃色即可。

心得

- 蓮藕須擦洗乾淨才去皮刨絲或切割，因中間空心處沾有之污泥是很難清除的。

Ingredients

225 g lotus root
100 g minced fish
2 Chinese mushrooms
2 tbsp dried shrimps
1 tbsp diced spring onion
1 egg yolk
1 tbsp plain flour
1/4 tsp salt

Seasonings

1/2 tsp salt
1/2 tsp sugar
dash of sesame oil
pinch of pepper

Method

1. Scrape lotus root to remove skin, grate into long shreds. Soak and dice mushrooms and dried shrimps separately and saute in a little oil. Dish.
2. Season minced fish, mix well with the rest of ingredients, divide into 12 equal portions.
3. Add 2 tbsp of oil in heated pan, spoon in each portion of lotus root mixture, press flat the mixture to form a round patty, shallow-fry till golden brown on both sides. Dish and serve.

Practical Tips

- Scrub lotus root thoroughly before scraping, any dirty mud trapped in holes of lotus root is difficult to clean.

煎釀三寶
Fried Stuffed Assorted Vegetables

材料

絞碎鯪魚肉 160 克
葱粒 1 湯匙
實豆腐 1 件
矮瓜（茄子）1 條
尖青椒 2 隻

調味料

鹽 1/2 茶匙
糖 1/4 茶匙
胡椒粉少許
生粉 1 湯匙
水 1 湯匙

豉汁

水 4 湯匙
豆豉茸 1 湯匙
糖 1/2 茶匙
生粉 1/2 茶匙
生抽 1 茶匙
麻油 1/2 茶匙

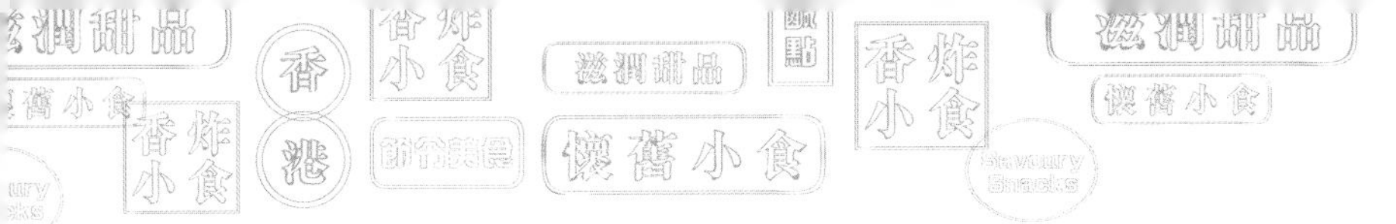

Ingredients

160 g minced dace
1 tbsp diced spring onion
1 piece firm bean curd
1 eggplant
2 green chillies

Seasonings

1/2 tsp salt
1/4 tsp sugar
shakes of pepper
1 tbsp cornstarch
1 tbsp water

Sauce mix

4 tbsp water
1 tbsp mashed fermented black beans
1/2 tsp sugar
1/2 tsp cornstarch
1 tsp light soy sauce
1/2 tsp sesame oil

做法

1. 絞碎鯪魚肉加入調味料，順一方向攪勻至起膠，加入葱粒拌勻候用。
2. 實豆腐切成三角形，矮瓜斜刀切厚塊，尖青椒去籽切塊。
3. 用小刀挑鯪魚肉餡一小份，分別釀入豆腐角、矮瓜及尖青椒中，置熱油中半煎炸至熟及呈金黃色，瀝油上碟。
4. 將豉汁用料混合，用小火煮滾，淋在煎釀三寶上，或另置小碟作蘸料用。

心得

- 建議三寶依次序落鑊，豆腐耐煎耐熱，可以先落鑊；其次是煎釀青椒、矮瓜，油溫要較高，可最後猛油落鑊，用高溫快煮法以保存其鮮艷色澤。

Method

1. Add seasonings into minced dace, stir well in one direction until quite stiff, add diced spring onion and mix well.
2. Cut firm bean curd into triangles, make a slit on the cut side; cut eggplant into thick slices; deseed and cut green chillies into segments.
3. Use a small knife to stuff in a spoonful of minced fish mixture on each of the assorted vegetables. Shape well and shallow-fry until cooked and golden brown. Drain and dish up.
4. Mix sauce ingredients together, cook in a low heat until fragrant, pour on top of stuffed vegetables or serve separate as a dip.

Practical Tips

- Shallow-fry the stuffed vegetables in the following suggested sequence: bean curds, green chilli and then eggplants which requires a higher heat to prevent it from discolouring.

生煎鍋貼
Wortips (Meat Dumplings)

份量：12 個 / Makes 12 pieces

材料

麵粉 150 克
鹽少許
豬油 1/2 湯匙
暖水約 125 毫升
絞腩肉 225 克
葱粒 1 湯匙
上湯 250 毫升

調味料

鹽 1/2 茶匙
糖 1/2 茶匙
生粉 2 茶匙
生抽 2 茶匙
酒 1/2 茶匙
胡椒粉、麻油各少許

做法

1. 麵粉過篩，加鹽及豬油，用適量暖水開成一軟滑粉糰，放置一旁待約 30 分鐘。
2. 腩肉加入調味料及葱粒，拌至起膠，置雪櫃內冷藏 30 分鐘。
3. 案板上灑粉，將粉糰分成 12 小份，每份用木棍開成圓塊圖 1~2，包入適量餡料，一邊埋口，一邊打摺成鍋貼形，先打摺後包入餡料圖 3，埋口圖 4。
4. 燒熱平底鑊，加入少許油，放入鍋貼，注入上湯 圖 5，加蓋文火煮約 15 分鐘至收汁。
5. 打開蓋，另加油 1 湯匙，煎至鍋貼底部呈金黃及鬆脆即可，熱食。

心得

- 將鍋貼餃打摺，需練習多次才成功。初做時，為免影響賣相，可用皮作餡，練習至合心水為止，重複多次，很快便可成為小師傅了！

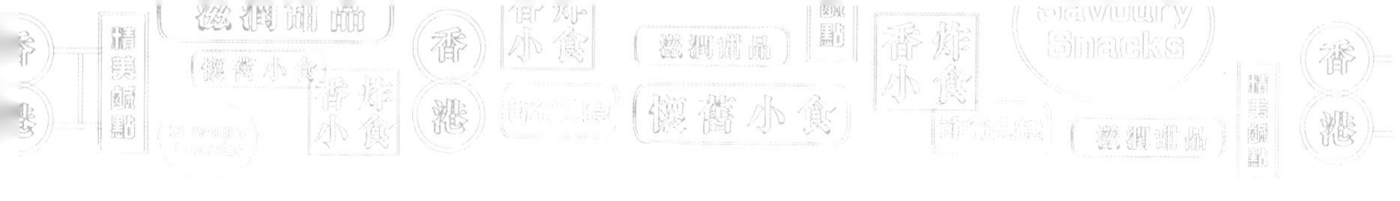

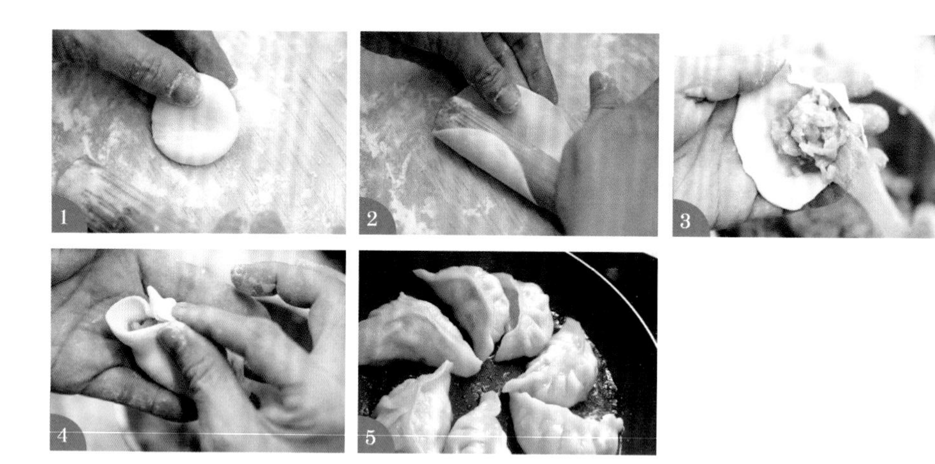

Ingredients

150 g plain flour
pinch of salt
1/2 tbsp lard
125 ml warm water
225 g minced pork belly
1 tbsp diced spring onion
250 ml chicken stock

Seasonings

1/2 tsp salt
1/2 tsp sugar
2 tsp tapioca starch
2 tsp light soy sauce
1/2 tsp Chinese wine
pinch of pepper
dash of sesame oil

Method

1. Sieve plain flour, add salt and lard. Mix in sufficient warm water to form soft smooth dough, rest for 30 minutes.
2. Season minced pork, add diced spring onion, stir until sticky. Chill in the refrigerator for 30 minutes.
3. Knead and divide dough into 12 equal portions on a floured board. Roll each piece of dough to a round Pictures 1~2 , wrap in a good lump of filling Picture 3. Fold, pleat and seal and pleat to the shape of wortips. Picture 4
4. Heat and grease a frying pan, add wortips and chicken stock Picture 5, cover and cook for 15 minutes until cooked.
5. Open the lid, add 1 tbsp of oil, shallow-fry the bottom of wortips till golden brown and crispy. Serve hot.

Practical Tips

- Shaping a pleated dumplings needs practice. For beginners, pinch a piece of dough for filling. Practice wrapping and pleating for several times and soon you will find yourself a professional dough chef!

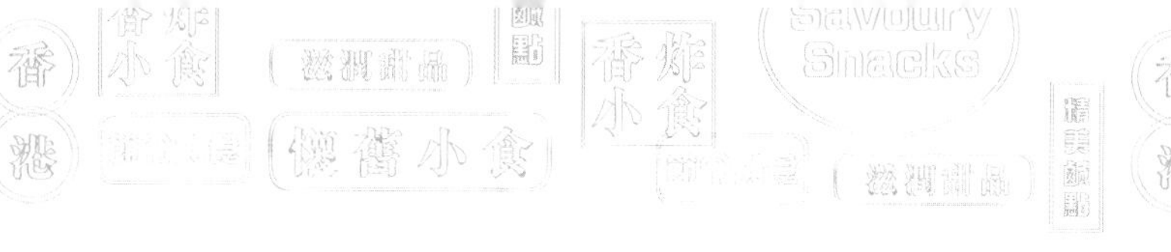

豬皮魷魚蘿蔔
Soy Pig's Skin, Squid and Turnips

豬皮蘿蔔乃是道地的香港街頭小吃，是「掃街」的必吃之選。

材料

已泡發豬皮 300 克
已泡發魷魚 1 隻
白蘿蔔 400 克
薑 1 片
蒜肉 1 粒
八角 1 粒

調味料

水 4 量杯
生抽 1/2 量杯
老抽 1/4 量杯
糖 1 湯匙
鹽 1/2 茶匙
麻油少許
胡椒粉少許

做法

1. 豬皮切塊，魷魚切件，白蘿蔔去皮切角。燒熱一鑊水，加入薑片，分次將豬皮、白蘿蔔及魷魚出水，瀝乾候用。
2. 起鑊燒油 2 湯匙，爆香蒜肉，加入白蘿蔔炒片刻，灒酒幾滴，加入八角及調味料煮滾，改用文火煮 15-20 分鐘至蘿蔔熟透。
3. 加入豬皮再煮 10 分鐘，最後加入魷魚，煮滾後離火加蓋，焗至入味，隨時享用。

心得

- 已泡發豬皮及魷魚於豆腐檔有售。
- 起鑊時可隨喜好拌入適量咖喱醬或豆瓣醬。
- 魷魚容易過火變硬，必須最後才加入，浸泡至入味。
- 可以灼熟生菜伴碟。

Ingredients

300 g blanched pig's skin
1 blanched squid
400 g turnips
1 slice ginger
1 clove garlic
1 piece star aniseed

Seasonings

4 cups water
1/2 cup light soy sauce
1/4 cup dark soy sauce
1 tbsp sugar
1/2 tsp salt
dash of sesame oil
shakes of pepper

Method

1. Trim pig's skin and squid into pieces, peel and make slanting cut for turnips. Blanch all ingredients separately in a pot of boiling water, add ginger slice. Rinse in cold water, drain.
2. Heat 2 tbsp of oil in wok, add garlic, stir fry turnips for a while, sizzle in wine, add star aniseed and sauce mix, bring to the boil, simmer for 15-20 minutes until cooked.
3. Add blanched pig's skin, cook for another 10 minutes. Add squid, turn off heat when re-boils. Cover and soak until ready to serve.

Practical Tips

- Blanched pig's skin and squid is sold at bean curd shops in local wet market.
- Can start with an adequate amount of curry paste or broad bean paste before adding turnips into the wok as desired.
- Squid tends to become overcooked easily, must add last before serving.
- Can serve this snack with blanched lettuce.

煎薄罉
Chinese Pancakes

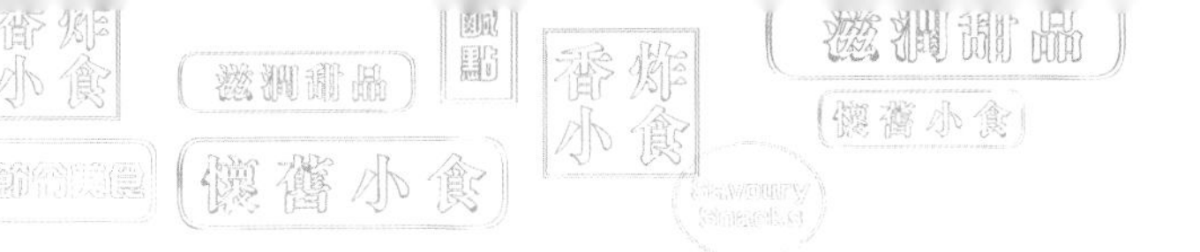

材料

麵粉 100 克
生粉 20 克
水 200 毫升
叉燒 60 克
冬菇 2 朵
蝦米 2 湯匙
韭菜 20 克

調味料

鹽 1/2 茶匙
糖 1/4 茶匙
油 1 茶匙
胡椒粉少許
麻油少許

做法

1. 麵粉及生粉同篩，加水調勻成薄漿，加入調味料，拌勻。
2. 叉燒切碎；冬菇浸軟，去蒂切粒；蝦米浸軟切碎；韭菜洗淨切粒。將所有材料加入粉料中拌勻。
3. 平底鑊燒油，分次加入麵漿使鋪滿鑊底，煎至金黃色，反轉再煎，中途可加油，半煎炸至鬆脆。
4. 上碟，熱食。

心得

- 煎薄罉時可隨喜好調校薄罉的份量，厚薄隨意。
- 反轉後，中途加一點油可增加薄餅的鬆脆口感。

Ingredients

100 g plain flour
20 g tapioca starch
200 ml water
60 g roast pork
2 Chinese mushrooms
2 tbsp dried shrimps
20 g green chives

Seasonings

1/2 tsp salt
1/4 tsp sugar
1 tsp oil
shakes of pepper
dash of sesame oil

Method

1. Sieve plain flour and tapioca starch together, add water, mix well to form a thin batter, add seasonings.
2. Dice roast pork; soak and remove stalk of Chinese mushrooms, dice; soak and dice dried shrimps; clean and dice green chives. Put all the ingredients in the batter, mix well.
3. Heat oil in frying pan, spoon in sufficient batter mixture to cover the bottom of pan. Shallow-fry till golden brown on both sides.
4. Dish and serve.

Practical Tips

- You can adjust the thickness of pancakes as desired.
- Add oil in between frying to enhance crispiness of pancakes.

香茅雞中翼
Lemongrass Chicken Wings

材料

雞中翼 8-10 隻
香茅 1/2 枝
香茅粉 1 湯匙

醃料

鹽 1/2 茶匙
糖 1 茶匙
酒 1 茶匙
生粉 1 湯匙
油 1 茶匙

做法

1. 雞中翼洗淨，吸乾水分，均勻地抹上香茅粉。
2. 香茅拍扁，斜刀切碎。
3. 將醃料拌勻，用手將醃料與雞翼拌勻，加入香茅碎，醃 5-6 小時。
4. 大火將雞中翼蒸 8 分鐘至八成熟，取出除去香茅碎，待冷卻片刻，用平底鑊加少許油，以中火將雞翼煎至熟透及呈金黃色，上碟，熱食。

心得

- 用手拌入醃料，雞翼較易入味。
- 將雞翼攤凍才下油鑊，可避免濺油溢出。

Ingredients

8-10 chicken wings, middle joint
1/2 stick lemongrass
1 tbsp lemongrass powder

Marinade

1/2 tsp salt
1 tsp sugar
1 tsp wine
1 tbsp cornstarch
1 tsp oil

Method

1. Clean and dry chicken wings, rub in lemongrass powder.
2. Crush and cut lemongrass into pieces.
3. Prepare marinade, rub to mix well with the chicken wings, marinate for 5-6 hours.
4. Steam chicken wings over medium heat for about 8 minutes until almost cooked, remove bits of lemongrass, cool. Pan fry with a little oil until well cooked and golden brown in colour, dish and serve hot.

Practical Tips

- Rub chicken and marinade together with hands helps to enhance flavour.
- Leave the steamed chicken wings to cool before frying can prevent the oil from splashing.

齋滷味
Gluten Snacks

材料

炸麵筋 275 克
麻油 1/2 湯匙
麥芽糖 1 湯匙

甜酸料

茄汁 3 湯匙
白醋 3 湯匙
素上湯 250 毫升
糖 2 湯匙
鹽 1/4 茶匙

生粉水

生粉 1 茶匙
水 1 湯匙
（調勻）

做法

1. 將炸麵筋炸硬，撈起瀝乾油分。
2. 鑊中留少許油，加入甜酸料，煮滾，放入炸過之麵筋，慢火煮至入味，約 15-20 分鐘。
3. 最後加入麻油及麥芽糖炒片刻，用生粉水埋芡即可。

心得

- 購買回來的炸麵筋雖然已經炸過，但必須再翻炸才炆煮，以免影響其質感。若怕炸麵筋太油膩，可在翻炸前用吸油紙吸去其多餘的油分，但切勿用水洗，因有濺油的危險。

Ingredients

275 g fried gluten
1/2 tbsp sesame oil
1 tbsp malt syrup

Sweet and Sour Sauce

3 tbsp tomato ketchup
3 tbsp rice vinegar
250 ml vegetable stock
2 tbsp sugar
1/4 tsp salt

Cornstarch Solution

1 tsp cornstarch
1 tbsp water
(mix well)

Method

1. Deep-fry the fried gluten until firm, drain.
2. Heat sweet and sour sauce until boils, add fried gluten, simmer for about 15-20 minutes.
3. Add sesame oil and malt syrup, stir for a while, thicken with cornstarch solution, dish and serve.

Practical Tips

- Deep-fry the fried gluten once more helps to keep the texture and shape after stewing. Do not rinse gluten, use kitchen paper to absorb excess grease for cleaning purpose.

滋潤甜品
Nourishing Desserts

無論那少甜、走甜的健康飲食概念已興起了一時，甜品，對於很多人，尤其是女士們來說，是無法抗拒的，包括我這個愛吃愛煮的。每餐過後，總愛找一口甜點往嘴裏送，才捨得離桌，才感到完美，那怕只是一小口的豆沙鍋餅、一小件的桂花糕，已令人感到一股心滿意足的溫暖。自家製甜點，可自由調校甜度，為自己及家人「度身訂造」；甜品如木瓜杏汁銀耳露、芝麻糊、鮮奶燉蛋等更有滋潤養顏定心的功效，“comfort food”原來衍生於此！

最末一章，以甜點代表一曲安歌。生活如歌，以美食奉獻，以書會友，希望你們都喜歡！

豆腐花
Soy Bean Custard

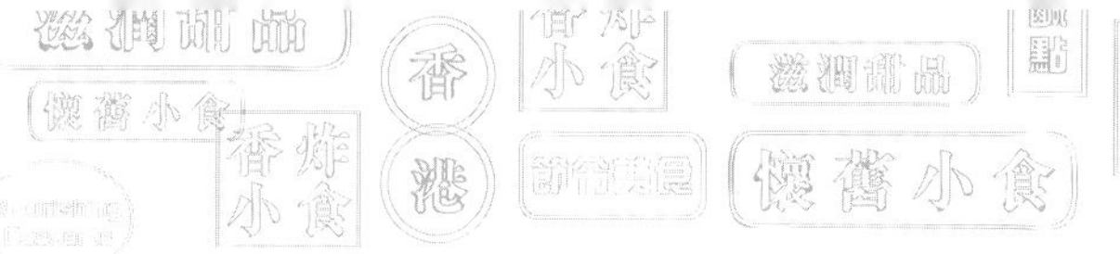

豆腐花
Soy Bean Custard

材料

黃豆 250 克
清水 2 公升
粟粉 1 平湯匙
熟石膏粉 1 平茶匙
冷開水 125 毫升

Ingredients

250 g soy beans
2 litres water
1 level tbsp cornstarch
1 level tsp gypsum (edible)
125 ml drinking water

伴吃糖水

冰糖 200 克
水 250 毫升
或黃砂糖適量

To Serve (syrup)

200 g rock sugar
250 ml water
or serve with brown sugar

1

2

3

做法

1. 黃豆洗淨，浸 5-6 小時，去衣後瀝乾。分次放入攪拌機內，加水攪成豆漿。
2. 將豆漿連渣倒入大煲內煮滾至出味，需不停除掉泡沫及攪拌，以防黏底。圖 1
3. 離火，冷卻後隔去豆渣，取豆奶 8 杯。
4. 將粟粉及石膏粉與冷開水拌勻，置木桶或深碗內，將豆奶燒滾，迅速撞入石膏粉漿內，加蓋靜置 15-20 分鐘至凝固，即成豆腐花。圖 2~3
5. 冰糖用水煮溶，與豆腐花同上，或灑上黃砂糖，熱食。

心得

- 豆渣連水一同煮成豆漿時，要用慢火及不停攪拌，並要用較大的煲來煮，可防止溢瀉及必須不停除掉泡沫。喜歡的話，可加入薑汁同煮。
- 用濕布包裹平碟，用作蓋子，可預防形成倒汁水。

Method

1. Rinse soy beans and soak for 5-6 hours, remove skin, rinse and drain well. Liquidize with adequate amount of water in electric blender to form soy bean milk.
2. Bring to the boil, keep stirring to avoid sticking and keep removing foams on the surface. Picture 1
3. Remove from heat, cool and drain to get 8 cups soy bean milk.
4. Mix cornstarch and edible gypsum with drinking water, put in deep mixing bowl, bring soy bean milk back to the boil and quickly pour into gypsum solution, cover and stand for 15-20 minutes until set. Pictures 2~3
5. Dissolve rock sugar in water to form syrup, serve with soy bean custard. Or serve soy bean custard with brown sugar.

Practical Tips

- Use low heat in cooking soy bean milk and must keep stirring to avoid residue from sticking, use a larger saucepan to avoid soy bean milk from splashing over. Keep removing foam from time to time. Add ginger juice to taste as desired.
- To cover the hot bean curd custard, wrap a large plate in a damp tea towel to avoid the formation of water vapor which will affect the setting effect.

芒果布甸
Mango Pudding

材料

芒果 2 個
芒果味啫喱粉 2 盒
雞蛋 2 個
滾水 250 毫升
鮮奶 450 毫升
蜜餞紅車厘子 2 粒（或免）
淡奶少許（或免）

Ingredients

2 mangoes
2 box mango flavour jelly
2 eggs
250 ml boiling water
450 ml fresh milk
2 candied red cherries (optional)
a little evaporated milk (optional)

做法

1. 芒果去皮、去核、起肉，用攪拌機將果肉打成茸。
2. 啫喱粉與雞蛋混和，攪至糖溶，徐徐拌入攝氏 90 度的滾水，不停攪拌至啫喱粉溶透為止，加入鮮奶及芒果茸，拌勻。
3. 倒入啫喱杯內，置雪櫃內雪至凝固，用車厘子裝飾，隨意加淡奶伴吃。

心得

- 芒果布甸可原杯上桌，或自模型倒出：從雪櫃取出布甸，靜置室溫一會，原杯浸一浸熱水，蓋上小碟，倒扣輕搖一下，布甸「卟」的一聲發響，即已成功脫模。

Method

1. Peel and stone mango, take flesh, blend to form mango puree.
2. Mix mango jelly with beaten eggs, mix well to dissolve the sugary crystals, gradually add in boiling water (90°C), keep stirring to make sure the gelatine is dissolved thoroughly, add fresh milk and mango puree.
3. Mix well and fill in jelly moulds, chill until set, unmould and decorate with red cherries. Serve cold with evaporated milk.

Practical Tips

- To unmould the pudding, take out pudding and stand in room temperature for a short while, dip the mould in hot water for a second, cover with the serving plate, turn upside down and give it a brisk shake, pudding will be successfully unmoulded when a "pop" sound is heard.

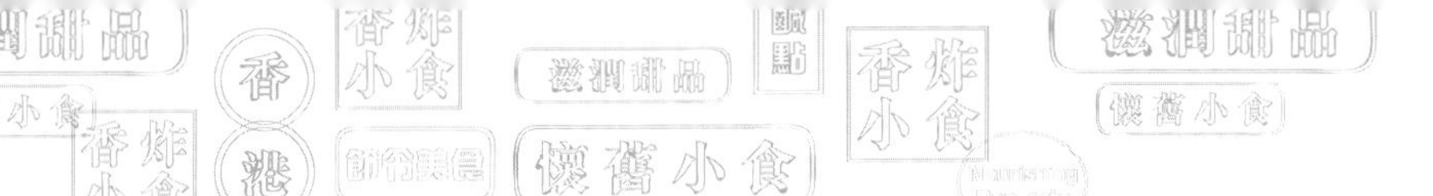

馬豆糕
Split Pea Pudding

份量：20 件 / Makes 20 pieces

材料

馬豆 50 克
粟粉 75 克
椰汁 250 毫升
鮮奶 200 毫升
大菜 15 克
水 1 公升
砂糖 100 克

Ingredients

50 g yellow split peas
75 g cornstarch
250 ml coconut milk
200 ml fresh milk
15 g agar agar
1 litre water
100 g castor sugar

做法

1. 預備方形糕盤，用冷水搪過。
2. 馬豆用慢火焓腍，過冷水瀝乾候用。
3. 粟粉用椰汁及鮮奶調勻。
4. 大菜放入 1 公升滾水內煮溶，過濾，離火加入砂糖及奶糊，開火攪至濃稠，滾後加入馬豆。
5. 將奶糊倒入方形糕盤內，待冷雪凍，倒出切件冷吃。

心得

- 用紅豆代替馬豆，便成為另一款糕品了。煮紅豆之水可留作煮大菜用。

Method

1. Rinse square tin with cold water.
2. Simmer yellow split peas until soft. Rinse and drain.
3. Mix cornstarch with coconut milk and fresh milk.
4. Dissolve agar-agar in 1 litre boiling water, drain, stir in sugar and milk solution. When thickens, add yellow split peas.
5. Pour mixture in square tin, cool and chill. Unmould, slice and serve cold.

Practical Tips

- For variations, use cooked red beans instead of yellow split peas for a change of taste.

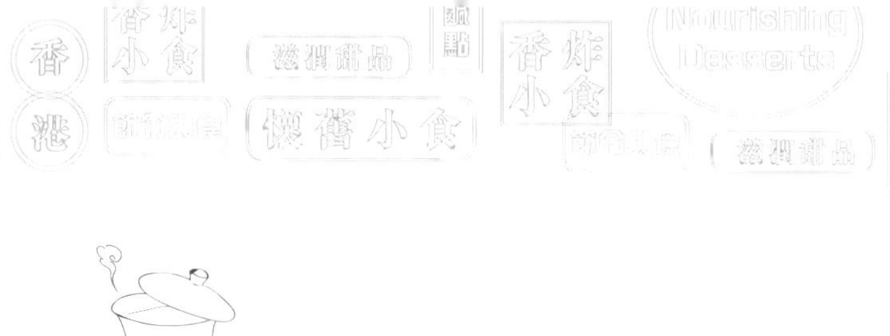

紅豆糕
Red Bean Pudding

份量：10 件 / Makes 10 pieces

材料

紅豆 75 克
魚膠粉 2 滿湯匙
砂糖 6 湯匙
滾水 300 毫升

做法

1. 紅豆浸透，置深碗內，注入浸過面之滾水，大火燉 1 小時至豆腍為止。隔去多餘水分，待涼。
2. 魚膠粉與砂糖拌勻，將魚膠糖水分成三份，注入滾水 300 毫升，不停攪拌至魚膠粉溶透。
3. 預備方形膠盒，將魚膠糖水分成三份，注入一層魚膠糖水，約 1/2 厘米厚，置雪櫃內冷藏 15-20 分鐘至凝固。
4. 將 1 份魚膠糖水加入已燉好之紅豆內拌勻，鋪在已凝固之啫喱上，待紅豆餡冷藏至凝固後再加一層透明的魚膠糖水，做成三層的紅豆糕，置雪櫃內冷藏 4-5 小時。
5. 取出紅豆糕，用利刀切件冷吃。

心得

- 紅豆隔水燉腍可保持原粒狀；將紅豆煲熟較省時但易爆裂。

Ingredients

75 g red beans
2 tbsp gelatine
6 tbsp castor sugar
300 ml boiling water

Method

1. Soak red beans for several hours, drain and put into a deep bowl. Add boiling water to cover the beans and steam over high heat for 1 hour until tender. Drain excessive water. Let cool.
2. Mix gelatine with sugar. Pour in 300 ml boiling water. Stir until gelatine is well dissolved.
3. Divide gelatine solution into three portions. Pour a layer of gelatine solution in a plastic box to 1/2 cm thick. Chill for or about 15-20 minutes until set.
4. Mix one portion of gelatine solution with the cooked red beans. Spread on top of set jelly. Chill the red beans layer until set. Pour the remaining gelatin solution on set red beans and chill for 4-5 hours for a layered result.
5. Unmould, cut into pieces with a sharp knife, serve cold.

Practical Tips

- Steaming helps to keep beans in shape. Boiling saves cooking time but the red beans will become too soft.

桂花糕
Osmanthus Jelly

份量：6-8 件 / Makes 6-8 pieces

材料	Ingredients
乾桂花 8-10 克	8-10 g dehydrated osmanthus
魚膠粉 20 克	20 g gelatine
冰糖 80-100 克	80-100 g rock sugar
熱水 500 毫升	500 ml hot water

做法

1. 桂花用熱水浸泡 15 分鐘後隔清。
2. 冰糖拍碎，與桂花茶一同煮至溶解成桂花糖水，待涼片刻。
3. 用鐵匙灑入魚膠粉，輕手攪動至魚膠粉完全溶透。將魚膠桂花糖水注入方形膠盒內約 5 厘米深。
4. 適量地加入泡過的桂花，間中攪動，勿讓桂花沉於膠盒底部。也可將膠盒置於冰水及冰塊上，不停攪動（5-8 分鐘）至魚膠桂花糖水凝結成糕狀。
5. 桂花糕置雪櫃內冷藏半天或過夜；取出，反扣離盒，切成小件，即成。

心得

- 魚膠粉不宜直接加火溶解，以免黏作一團；宜隔着蒸氣及必須用鐵匙攪拌魚膠粉，使它逐漸地溶透。看附圖
- 冷凍糕點在離模前，宜坐入熱水中浸 1-2 秒，方便脫模。
- 甜度及桂花的濃度可自由調校。

Method

1. Soak dehydrated osmanthus in hot water for 15 minutes until fragrant. Drain.
2. Crush rock sugar, add into osmanthus tea and boil until rock sugar is dissolved. Cool down a bit.
3. Sprinkle gelatine into osmanthus syrup with a metal spoon, stirring slightly until gelatine is well dissolved. Pour into oblong plastic box to 5 cm deep.
4. Add adequate amount of the blanched osmanthus, stir occasionally to avoid osmanthus from sinking to the bottom. You can also put the box in a basin of ice water with ice cube, keep stirring (5-8 minutes) to keep osmanthus floating in the jelly until set.
5. Chill in fridge for half day or overnight. Talk out, unmould and cut osmanthus jelly into pieces. Serve chilled.

Practical Tips

- Do not heat gelatine on direct heat to avoid lumps. Stirring over a steam of hot water with a metal spoon helps to dissolve gradually for a clear gelatine solution.
- Dip the box of Jelly in hot water for 1-2 seconds for easy unmoulding. As shown
- Adjust sweetness and fragrancy of osmanthus as desired.

蔗汁糕
Sugar Cane Pudding

材料

馬蹄粉 100 克
蔗汁 500 毫升
片糖 50 克
油 1/2 湯匙

Ingredients

100 g water chestunt flour
500 ml sugar cane juice
50 g slab sugar
1/2 tbsp oil

做法

1. 馬蹄粉置深碗內，加入 250 毫升蔗汁，拌勻，浸片刻。
2. 片糖切碎，加入 250 毫升蔗汁內慢火煮溶，傾入馬蹄粉蔗汁水內拌勻，過濾。
3. 將混合物倒入鍋內，加油，邊攪邊煮至半透明及成厚糊狀，倒入已塗油的鐵模內，大火蒸 15 分鐘。
4. 取出，冷卻片刻，切件，冷熱吃皆可。

心得

- 馬蹄粉水必須過濾，去除雜質。
- 馬蹄粉糊入鐵模後，用掃了油的鐵匙趁熱將糕面掃至平滑。

Method

1. Mix water chestnut flour with 250 ml sugar cane juice in mixing bowl, soak for a while.
2. Chop slab sugar, dissolve in 250 ml sugar cane juice under low heat, pour into water chestnut flour solution, drain.
3. Heat mixture in saucepan, add oil, keep stirring until transparent and thickened, pour into greased tin. Steam over high heat for 15 minutes.
4. Remove from heat, cool and cut into pieces, serve warm or cold.

Practical Tips

- Drain water chestnut solution well to remove coarse lumps.
- Pour thickened pudding into greased tin and smoothen the surface with a greased metal spoon before steaming.

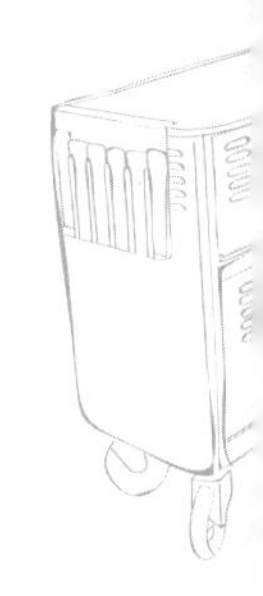

椰汁糕
Coconut Pudding

份量：20 小件 / Makes 20 pieces

材料	Ingredients
水 250 毫升	250 ml water
砂糖 125 克	125 g castor sugar
魚膠粉 25 克	25 g gelatine
椰汁 250 毫升	250 ml coconut milk
鮮奶 250 毫升	250 ml fresh milk
蛋白 150 毫升	150 ml egg white

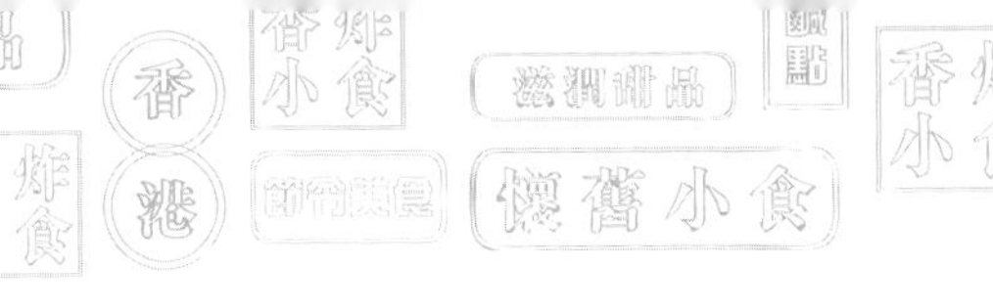

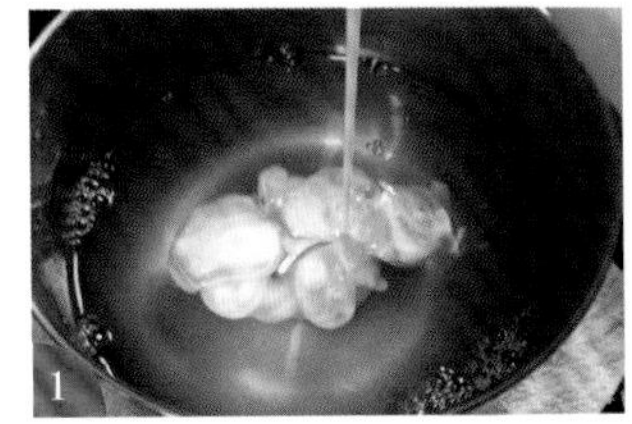
1

2

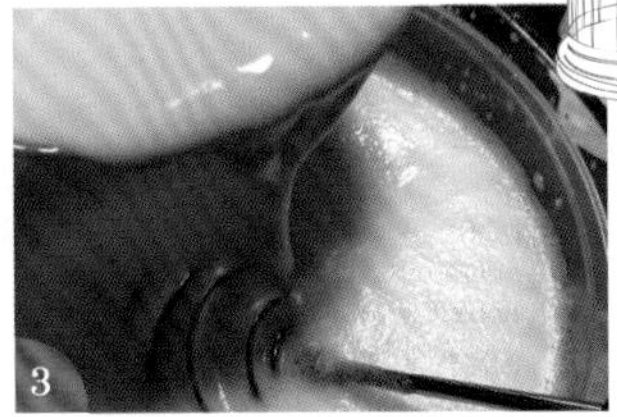
3

做法

1. 燒滾水 250 毫升，加入砂糖煮溶，離火待暖，輕拌入魚膠粉，攪至溶透，加入椰汁及鮮奶 圖 1，置雪櫃內凍至半凝固如厚忌廉狀。
2. 蛋白置深碗內打至企身 圖 2，逐少加入半凝固的魚膠奶糊 圖 3，打至材料完全混合。
3. 將混合物迅速倒入方形膠盒內，凍至凝固後倒出，切件冷吃。

心得

- 如冷藏魚膠奶糊過久而已凝固成固體，可將它坐在熱水上，便可回復流質狀，再重新冷凍成忌廉狀即可。
- 打蛋白時，不要太早加入魚膠奶糊，以免椰汁糕呈層次狀。
- 將蛋白魚膠奶糊隔着冰塊上打勻，可加速混合。如使用電打蛋器，宜用中慢速。

Method

1. Boil water, add sugar, when dissolved, cool down a bit, add gelatine, stir well, add coconut milk and fresh milk Picture 1, chill in the refrigerator until half set (like the consistency of thick cream).
2. Whip egg white in a deep bowl until stiff Picture 2, gradually whip in gelatine mixture until ingredients are well mixed Picture 3.
3. Immediately pour mixture in rectangular plastic box, chill until set, cut into pieces, serve cold.

Practical Tips

- If gelatine mixture is over-set, melt over steaming water and re-chill to get a creamy consistency.
- Never pour gelatine mixture into egg white until egg white is stiff enough to hold the mixture.
- Add egg white gelatine mixture gradually and beat on top of ice cubes to speed up blending. Use medium speed for blending the mixture if electric beater is used.

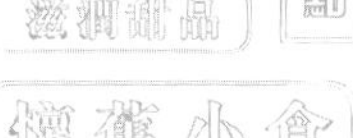

鮮奶燉蛋
Steamed Milk Custard

份量：4 碗 / Makes 4 bowls

材料

蛋液 250 毫升
水 250 毫升
鮮奶 250 毫升
冰糖 100 克

Ingredients

250 ml beaten egg
250 ml water
250 ml fresh milk
100 g rock sugar

做法

1. 將鮮奶逐少拌入蛋液中。
2. 冰糖用水煮溶，待涼，加入蛋料中過濾一次。
3. 將蛋料平均放入飯碗中，放入蒸籠慢火蒸至凝固（約 25 分鐘），熱食。

心得

- 如用鑊蒸，可蓋上保鮮紙隔去水氣，需時約 15-20 分鐘。
- 可隨喜好加入薑汁、杏仁汁或朱古力溶液。

Method

1. Add fresh milk into beaten egg gradually.
2. Dissolve rock sugar in 1 cup of water. Cool, gradually mix into the egg mixture and drain.
3. Divide mixture into 4 rice bowls, steam over gentle heat in a bamboo steamer for about 25 minutes or until set, serve hot.

Practical Tips

- If a metal steamer or wok is used, cover custard with glad wrap to avoid water vapour. Cooking time is about 15-20 minutes.
- Variations: Add ginger juice, almond sauce or melted chocolate to egg mixture as desired.

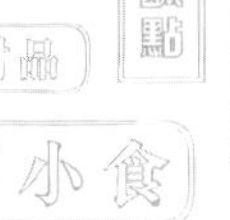

馬蹄露
Water Chestnut in Syrup

份量：10 碗 / Makes 10 bowls

材料

去皮馬蹄 450 克
清水 1 公升
冰糖 180 克
馬蹄粉 2 湯匙
雞蛋 2 個（打勻）

Ingredients

450 g peeled water chestnuts
1 litre water
180 g rock sugar
2 tbsp water chestnut flour
2 beaten eggs

做法

1. 馬蹄磨成茸，或拍扁剁碎。
2. 冰糖加水煮溶，加入馬蹄茸煮滾，除去泡沫。
3. 馬蹄粉加水 2 湯匙開溶，加入糖水中煮至微稠。
4. 離火打入蛋液，拌成蛋花，冷熱吃皆可。

心得

- 如想加入開邊綠豆，可預先將豆浸透，用大火蒸熟，約 20 分鐘，取出與馬蹄茸一同放入糖水中便可。

Method

1. Grate or crush and chop water chestnuts.
2. Bring water and rock sugar to boil, add water chestnuts, remove foam when re-boils.
3. Mix water chestnut flour with 2 tbsp of water. Add to syrup, stir till thickened.
4. Remove from heat, stir in beaten eggs. Serve either hot or cold.

Practical Tips

- Some people like to add green split beans in the syrup. Soak, drain and steam the split beans over high heat for about 20 minutes or until cooked, add into boiling syrup, stir well and serve.

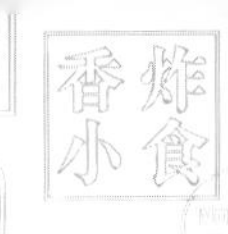

香芋西米露
Taro in Sago Syrup

份量：12 碗 / Makes 12 bowls

材料

荔浦芋 275 克
西米 100 克
冰糖 150 克
水 725 毫升
淡奶 125 毫升
椰汁 150 毫升

做法

1. 荔浦芋去皮後切粒，用大火蒸熟，趁熱壓成芋茸，或原粒使用。
2. 西米用大滾水浸 15 分鐘，間中攪拌避免黏作一團，瀝乾後置滾水內煮至半透明，瀝乾，沖冷水候用。
3. 冰糖加水煮滾，加入已沖透之西米及芋茸，煮滾，離火伴入椰汁及淡奶，趁熱進食。

心得

- 西米只需煮至半透明便可離火，一過冷水便會呈透明狀。

Ingredients

275 g taro
100 g sago
150 g rock sugar
725 ml water
125 ml evaporated milk
150 ml coconut milk

Method

1. Peel and dice taro, steam over high heat until cooked. Mash while still hot or just serve in cubes.
2. Soak sago in boiling water for 15 minutes, stir occasionally. Drain and boil until almost cooked, keep stirring. Drain, rinse under tap water and drain again.
3. Dissolve rock sugar in water, add sago and mashed taro, stir until re-boils. Remove from heat, stir in evaporated milk and coconut milk. Serve hot.

Practical Tips

- Start rinsing semi-transparent sago under tap water when it will turn to transparent after rinsing. Drain before use.

木瓜杏汁銀耳露
Papaya and White Fungus in Almond Sweet Tea

材料

木瓜 100 克（切粒）
杏仁粉 6 湯匙
鮮奶 100 毫升
雪耳 1 個
冰糖 100 克
薑 1 片
水 750 毫升

Ingredients

100 g papaya, diced
6 tbsp ground almond
100 ml fresh milk
1 piece white fungus
100 g rock sugar
1 slice ginger
750 ml water

做法

1. 杏仁粉用鮮奶開勻。
2. 雪耳浸軟去蒂，切成小朵，出水瀝乾。
3. 燒滾水，加薑片及雪耳，用小火煮腍，加入冰糖，煮溶。拌入杏仁奶漿，用小火煮 2-3 分鐘，試味。
4. 離火加入木瓜粒；冷熱吃皆可。

心得

- 離火後，可隨意立即伴入蛋白，增加口感。

Method

1. Blend ground almond with fresh milk.
2. Soak white fungus until soft, remove stalk and cut into small florets, blanch and drain well.
3. Bring water to the boil, add ginger slice and white fungus, boil until soft and cooked, add rock sugar, stir to dissolve. Add almond milk and cook for 2-3 minutes under low heat. Season to taste.
4. Remove from heat, add papaya cubes. Serve hot or cold.

Practical Tips

- As a variation, stir in beaten egg white right after tuning off heat.

腰果露
Cashew Nut Sweet Tea

份量：8 碗 / Makes 8 bowls

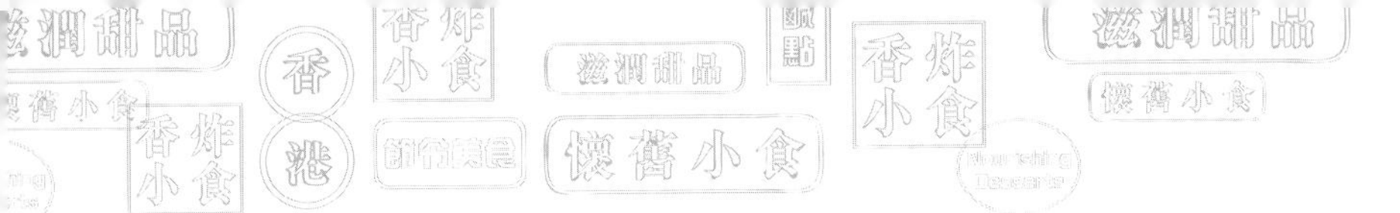

材料

腰果 200 克
鹽 1/2 茶匙
水 750 毫升（焓腰果用）
白米 1 湯匙
水 1.5 公升
冰糖 150 克

Ingredients

200 g cashew nuts
1/2 tsp salt and 750 ml water
(for boiling cashew nuts)
1 tbsp rice
1.5 litres water
150 g rock sugar

做法

1. 腰果原粒放入鹽水內焓 2 分鐘，取出，隔清水分，吹乾後放暖油內炸至浮起及呈微金黃色（以微爆開為合），撈起吸去多餘油分。
2. 白米用 125 毫升水浸透，約 1 小時，放入攪拌機攪成米漿，過濾。
3. 將炸好的腰果及 375 毫升水放入攪拌機攪成腰果漿。
4. 將餘下之 1 公升水煮滾，加入冰糖煮溶，加入腰果漿及適量米漿，煮至微稠即可供吃。

心得

- 炸腰果時不宜炸過火，否則糖水會帶燶味；用鹽水焓過腰果可去「青」味。煮腰果露時須不停攪拌，否則很易黏煲底。腰果含澱粉質較其他果仁高，不宜加太多米漿，也可隨意加入少量淡奶，但不宜太多，否則會太膩口。

Method

1. Boil cashew nuts in salt solution for 2 minutes, drain and air-dry. Deep-fry until golden brown. Remove and drain.
2. Soak rice in 125 ml of water for 1 hour. Blend to form rice solution, drain well.
3. Blend fried cashew nuts in 375 ml of water to get a fine solution.
4. Dissolve rock sugar in the remaining water, stir in cashew nut solution and thicken slightly with rice solution, mix well, serve hot.

Practical Tips

- Boiling cashew nuts in salt water helps to remove the raw flavour. Do not overcook cashew nuts to avoid a "burnt" taste in the sweet tea. Cashew nut contains a high percentage of starch, keep stirring to avoid sticking and do not over-thicken the dessert.

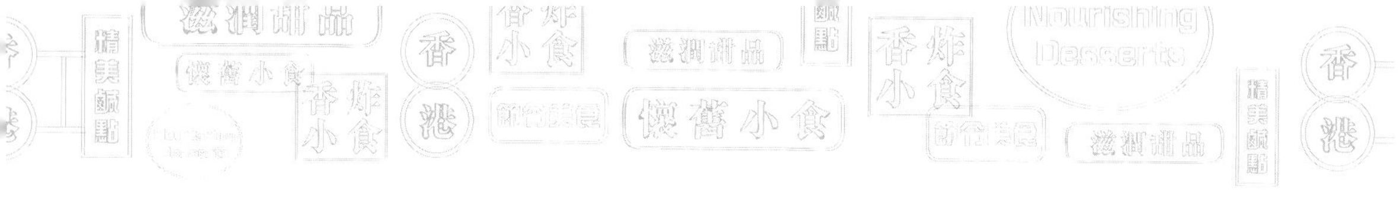

材料	Ingredients
黑芝麻 150 克	150 g black sesame
白米 4 湯匙	4 tbsp rice
冰糖 200 克	200 g rock sugar
水 1.5 公升	1.5 litres water

芝麻糊
Black Sesame Sweet Tea

份量：8 碗 / Makes 8 bowls

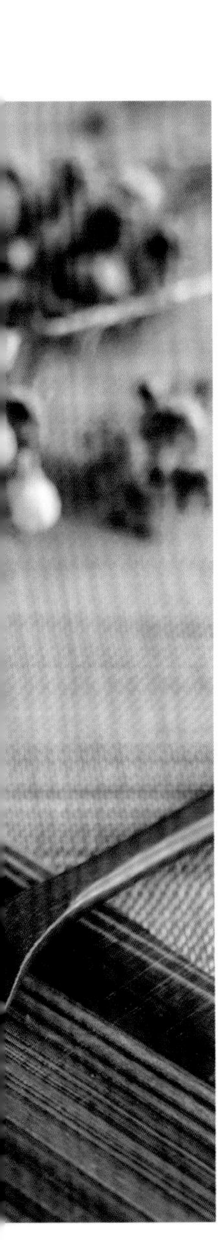

做法

1. 黑芝麻預早一天洗淨，去沙，瀝乾水分，吹乾後用白鑊炒香。
2. 白米洗淨，用 250 毫升水浸 1 小時，放入攪拌機內攪成米漿，過濾候用。
3. 黑芝麻用 375 毫升水攪至極幼，用密篩過濾成幼滑芝麻漿。
4. 將 875 毫升水燒滾，加入冰糖，煮至糖溶，加入芝麻漿，慢火煮片刻，拌入適量米漿，煮至稠。
5. 離火熱食。

心得

- 如隔渣用之篩不夠細密，可將隔出之芝麻漿再過濾一次，務求達至幼滑之效果。

Method

1. Rinse black sesame to clean, drain and air-dry overnight. Brown in a dry clean wok over medium heat.
2. Soak rice in 250 ml water for 1 hour, blend and drain to form a fine rice solution.
3. Blend black sesame in 375 ml water to form a fine solution, pass through a fine sieve.
4. Dissolve rock sugar in remaining water, add sesame solution, stir and cook for a while.
5. Serve hot.

Practical Tips

- If drainer or sieve is not fine enough, drain the sesame solution twice to get a smooth and fine result.

材料

去衣合桃 150 克
冰糖 200 克
白米 2 湯匙
水 1.25 公升
淡奶 50 毫升

Ingredients

150 g skinned walnuts
200 g rock sugar
2 tbsp rice
1.25 litres water
50 ml evaporated milk

合桃糊
Walnut Sweet Tea

份量：10 碗 / Makes 10 bowls

做法

1. 合桃放入淡鹽水烚 3 分鐘，洗淨瀝乾。待乾透後用慢火油炸至浮起及呈金黃色，取出隔去多餘油分。
2. 白米浸透後放攪拌機內加水少許打成漿狀，過濾成米漿。
3. 用 250 毫升水將合桃打成漿狀，過濾成合桃漿。
4. 將 1 公升水加冰糖煮溶，加入合桃漿及適量米漿煮至微稠。
5. 加入淡奶，拌勻熱食。

心得

- 去衣合桃可到海味店購買，有些雜貨店也有發售。去衣合桃不宜久存，最好貯在雪櫃內保鮮。

Method

1. Blanch skinned walnuts in salt water for 3 minutes. Drain and air-dry. Deep-fry over medium heat until cooked and brown. Drain.
2. Soak rice and liquidize in an electric blender, filter as rice solution.
3. Ground walnuts in 250 ml of water. Filter as walnut solution.
4. Dissolve rock sugar in remaining 1 litre of water, add walnut solution, when boils, thicken with rice solution.
5. Mix in evaporated milk, serve hot.

Practical Tips

- Shelled and skinned walnuts can be bought from most dried seafood shops. Keep skinned walnuts in refrigerator for longer shelf life.

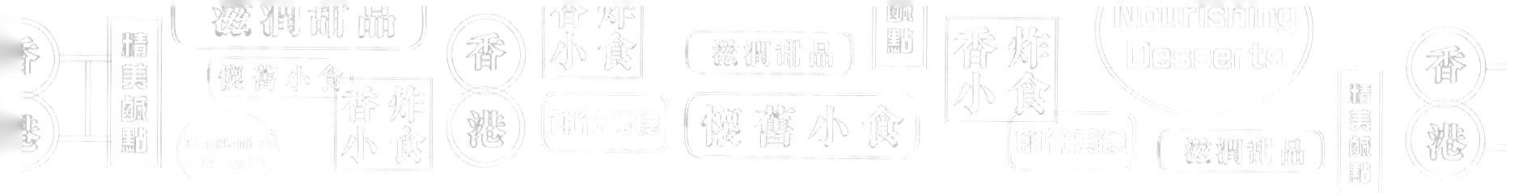

材料	Ingredients
南杏 100 克	100 g sweet apricot kernel
北杏 1 湯匙	1 tbsp apricot kernel
白米 2 湯匙	2 tbsp rice
水 1.25 公升	1.25 litres water
冰糖 150 克	150 g rock sugar
淡奶 50 毫升	50 ml evaporated milk

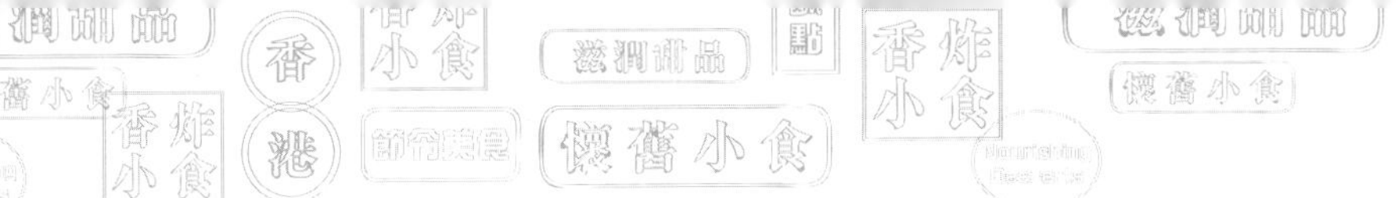

杏仁糊
Almond Sweet Tea

份量：8 碗 / Makes 8 bowls

做法

1. 南杏用熱水燙過，去衣，與北杏一起用 250 毫升水浸約 1 小時，放入攪拌機內打成杏仁漿，過濾候用。
2. 白米用 125 毫升水浸 1 小時，加入攪拌機內攪成米漿，過濾候用。
3. 用餘下之 875 毫升水將冰糖煮溶，加入杏仁漿，煮滾，用適量米漿調煮成糊狀，最後加入淡奶，拌勻熱食。

心得

- 杏仁化痰止咳，潤肺養顏，使皮膚潔白，幼滑。杏仁糊主要成分為連衣大南杏，北杏用多了帶苦澀味，藥性較重。甜度可自行調試，惟杏仁漿及米漿必須用密篩過濾，才達至幼滑之效果。

Method

1. Blanch sweet apricot kernel, remove skin, soak in 250 ml of water with apricot kernel for 1 hour, liquidize in an electric blender, filter as almond solution.
2. Rinse and soak rice in 125 ml of water for 1 hour, liquidize and filter as rice solution.
3. Boil the remaining 875 ml of water and dissolve rock sugar, add almond solution, stir until boils, thicken with rice solution, add evaporated milk, serve hot.

Practical Tips

- Chinese almonds help to relieve coughs, nourish the lungs and skin. The Almond Sweet Tea is specially good for ladies. To achieve a smooth and fine result, pass the almond and rice solutions through a fine sieve before use.

擂沙湯丸
Coated Sweet Dumplings

份量：12 個 / Makes 12 pieces

材料

糯米粉 175 克
粘米粉 25 克
豬油 1 湯匙
暖水 175 毫升
麻蓉 200 克
黃豆粉 2 杯，蘸料

Ingredients

175 g glutinous rice flour
25 g rice flour
1 tbsp lard
175 ml warm water
200 g sesame paste
2 cups soy bean powder, coating

做法

1. 糯米粉與粘米粉拌勻，加入豬油及暖水，拌成一軟滑粉糰，搓長，分成 12 小份。
2. 麻蓉搓長亦分成 12 小粒。
3. 將粉糰搓圓按扁，加入麻蓉，埋口搓圓，置大滾水內煮至湯丸浮起。
4. 撈起瀝乾水分，趁熱蘸上黃豆粉，熱食。

心得

- 黃豆粉在糕餅用料專門店或大型超級市場有售。

Method

1. Mix glutinous rice flour with rice flour, add lard and warm water to form a soft dough. Roll and divide into 12 equal portions.
2. Roll and divide sesame paste into 12 portions.
3. Roll and press the dough, stuff in sesame paste, seal and shape into a ball. Cook dumplings in fast boiling water.
4. Drain and coat with soy bean powder when still hot. Serve.

Practical Tips

- Soy bean powder can only be obtained in special bakery suppliers.

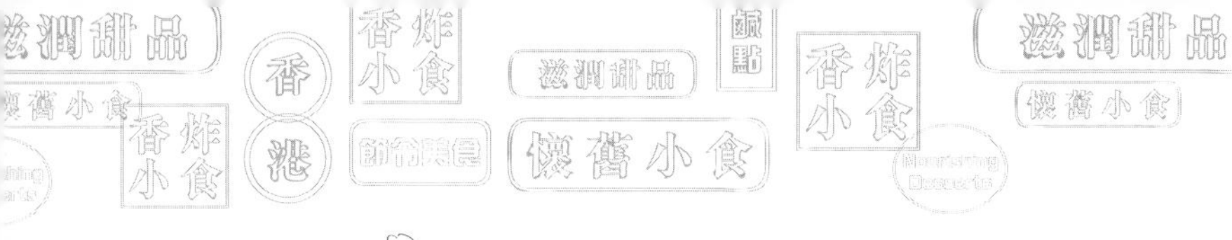

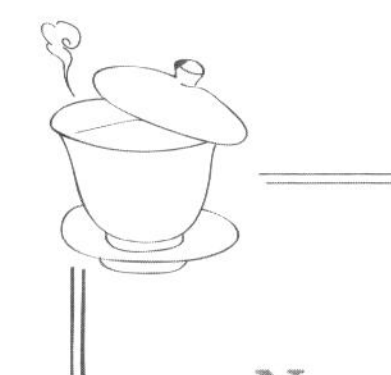

熱香餅
Nourishing Hot Cake

份量：8 件 / Makes 8 pieces

材料

自發粉 100 克
雞蛋 2 個
鮮奶 125 毫升
砂糖 1 湯匙

伴食

糖漿、牛油各少許

Ingredients

100 g self-raising flour
2 eggs
125 ml fresh milk
1 tbsp castor sugar

To Serve

golden syrup and butter

做法

1. 自發粉篩入大碗內，中間開穴，打入雞蛋及砂糖，逐少加入鮮奶，拌勻。
2. 隔去粉粒，置量杯內。
3. 平底鑊燒熱，加油少許，傾入適量麵漿，煎至兩面均呈金黃色。
4. 上碟，與牛油及糖漿同上。

心得

- 熱香餅以糖漿及牛油伴吃乃傳統吃法。喜歡的話，可煎兩塊大小一樣的熱香餅，中間夾一層豆沙，便成為「叮噹燒餅」了。

Method

1. Sieve self-raising flour in a deep bowl, make a well in the centre, add sugar and drop in eggs. Gradually mix in fresh milk, stir well.
2. Drain to remove lumps, keep batter in a jug.
3. Heat a frying pan till hot, add a little oil, pour in sufficient batter, turn and fry until golden brown on both sides.
4. Dish and serve hot cakes with golden syrup and butter.

Practical Tips

- For variations, sandwich two pieces of hot cakes with a layer of red bean paste as filling for a change of taste.

豆沙鍋餅
Red Bean Patties

份量：8-10 塊 / Makes 8-10 patties

材料	Ingredients
麵粉 150 克	150 g plain flour
雞蛋 2 個	2 eggs
鮮奶 375 毫升	375 ml fresh milk
豆沙 200 克	200 g red bean paste

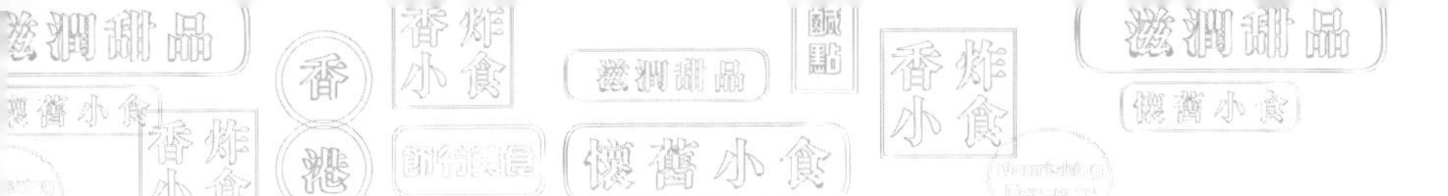

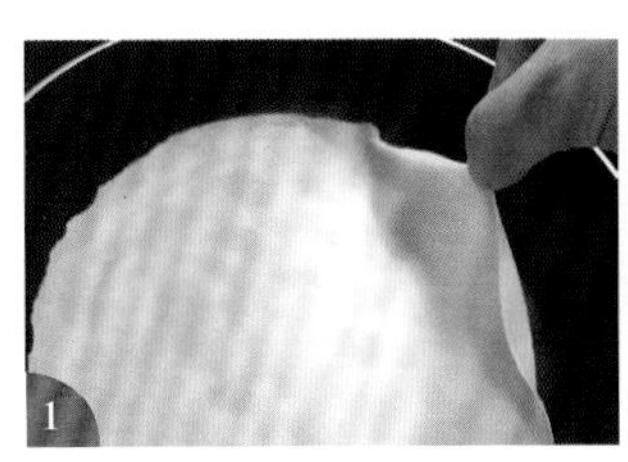

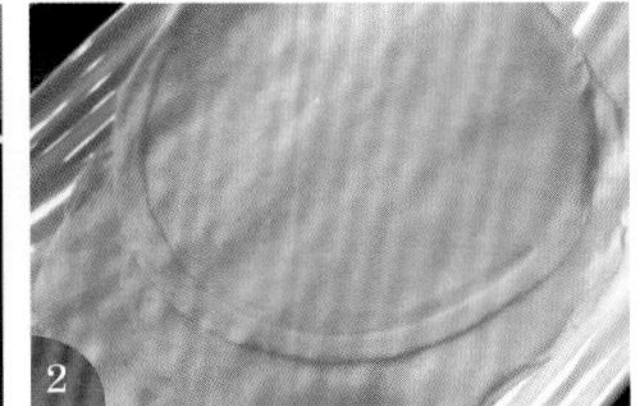

做法

1. 麵粉過篩，加雞蛋及鮮奶，開成稀滑奶漿，用平底易潔鑊煎成 8-10 塊薄餅，只需煎一面；留少許粉漿作黏口用。
2. 將薄餅鋪平，底向下，豆沙分成 8-10 份，逐份放上薄餅中央成長條形。
3. 自下方及兩邊餘位覆起，向前包成長扁狀，用少許粉漿黏口。
4. 燒熱油，用中火將豆沙鍋餅炸至金黃色，撈起瀝乾油分，切件趁熱享用。

心得

- 此薄餅可用作西式的「班戟」，佐以鮮果、忌廉、吞拿魚或果醬享用。圖 1~2

Method

1. Sieve flour, add eggs and fresh milk gradually, mix to form a smooth batter, fry into 8-10 pancake sheets. Fry on one side only (reserve a little batter for sealing pancakes).
2. Lie pancake sheet flat, underside downward, divide red bean paste into 8-10 portions. Spread each portion on centre of pancake sheet.
3. Fold pancake from lower flap and sides, roll to a flat oblong, damp edge with reserved batter, press to seal edges.
4. Heat the wok, deep-fry red bean patties in medium-hot oil until golden brown. Dish, drain, cut into sections. Serve hot.

Practical Tips

- The pancake sheet can also be served with fruit and fresh cream, tuna fish or jam. Pictures 1-2

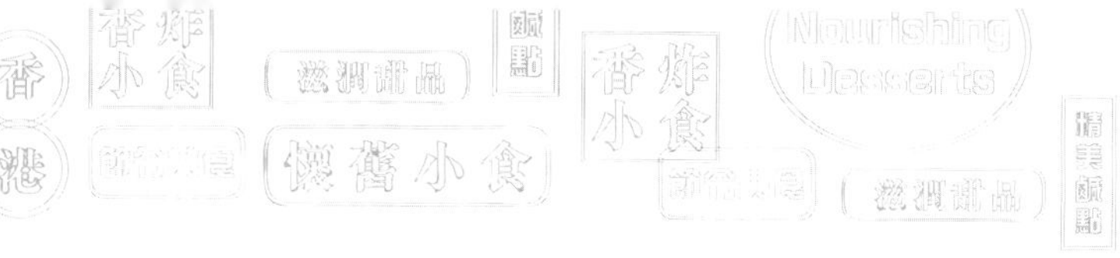

焗蓮蓉西米布甸
Baked Sago Pudding with Lotus Seed Puree

份量：3 小杯 / Makes 3 cups

材料

蓮蓉 75 克
西米 65 克
粟粉 1 湯匙
吉士粉 1/2 湯匙
椰汁 100 毫升
淡奶 2 湯匙
砂糖 50 克
牛油 15 克
水 250 毫升
蛋黃 1 個

Ingredients

75 g lotus seed puree
65 g sago
1 tbsp cornstarch
1/2 tbsp custard powder
100 ml coconut milk
2 tbsp evaporated milk
50 g castor sugar
15 g butter
250 ml water
1 egg yolk

做法

1. 預熱焗爐至 7 度（攝氏 210 度），耐熱小杯掃油。
2. 蓮蓉分成 3 份。
3. 西米用滾水浸 15 分鐘，間中攪拌，瀝乾，用滾水煮至半透明，過冷水，瀝乾。
4. 粟粉、吉士粉拌勻，加入椰汁及淡奶，拌成軟滑粉漿。
5. 燒滾 250 毫升水，加入西米、砂糖及牛油，拌勻；煮至西米呈透明狀，拌入粉漿，攪拌成糊狀，離火加入蛋黃，拌勻。加入小耐熱杯內至半滿，舀入一份蓮蓉，再加入 2 大湯匙西米糊至滿。
6. 將西米布甸放入焗爐內，焗 20 分鐘至面呈金黃色即成，熱食。

心得

- 西米用滾水浸洗，可保持西米的原粒狀。

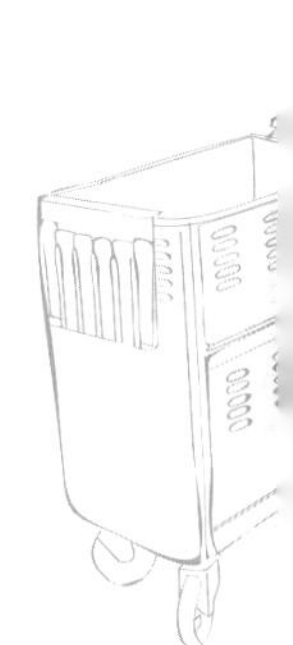

Method

1. Preheat oven to 210℃ , grease ovenproof basins.
2. Divide lotus seed puree into 3 portions.
3. Soak sago in boiling water for 15 minutes, stir occasionally, drain and cook in boiling water until semi-transparent, rinse under water tap, drain well.
4. Mix cornstarch and custard powder together, mix in coconut milk and evaporated milk, stir to form a smooth batter.
5. Boil 250 ml water, add sago, sugar and butter, keep stirring until sago looks transparent, stir in cornstarch batter, cook until thickened, remove from heat and stir in egg yolk. Mix well, spoon in mixture to half-fill oven proof basins, add a portion of lotus seed purée, fill up with 2 big spoonful of sago mixture.
6. Put sago puddings to bake for 20 minutes until golden brown, serve hot.

Practical Tips

- Soaking sago with boiling water helps to keep the sago in its shape.

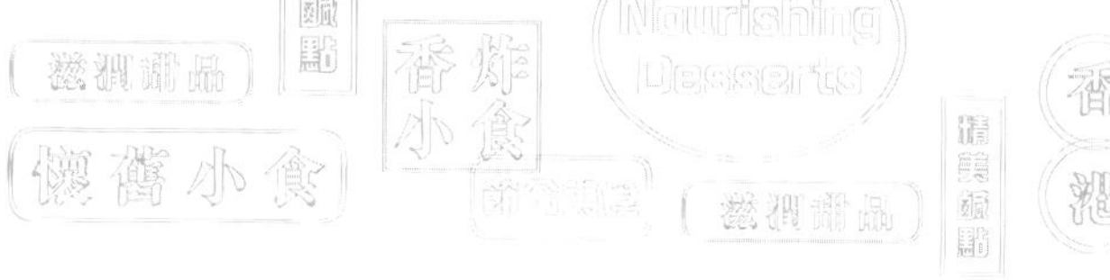

附錄

為了幫助讀者能輕易地成功製作書中介紹的八十八款小吃，除了要有基本的材料及精準的製法說明外，以下為各位提供一些關於糕點製作過程中必備的鬆化劑（Shortenings）及膨鬆劑（Raising Agents）的認識。

鬆化劑（Shortenings）

大部分在粉漿中、麵皮及糕餅（Batter, Dough and Pastries）上所用的油脂均有鬆化的作用，如菜油、花生油及豬油，而豬油更是常用的鬆化劑。鑑於家庭式製作糕餅時所需的份量不多，建議不妨自家貯備一點，放在冰箱，以作不時之需。

自製豬油

豬油製法

豬油膏少許（一般街市內相熟的肉檔會免費贈送），沖洗乾淨，吸乾水分後切成小塊，置乾淨的白鑊內，用小火略煎至出油，輕輕翻動肥膏至呈金黃色及鬆脆便可離火。待冷、隔渣後便可使用及貯存。

膨鬆劑（Raising Agents）

第一類
Natural Raising Agent

空氣 Air：屬於天然的膨鬆劑，將粉料過篩、攪拌材料或攪打雞蛋時所混入的空氣。蛋白中所含的蛋白"albumen"，當被攪打時會隨即與打入的氣體混在一起，而形成了一些膨鬆的空間，這也是氣體遇熱向上升的自然現象。

第二類
Chemical Raising Agents

泡打粉 Baking powder、
梳打粉 Bicarbonate of soda：
化學性的，遇熱時會釋放出二氧化碳，讓麵皮、糕餅自然膨鬆因而有鬆脆效果。

第三類
Biological Raising Agent

酵母 Yeast：屬生物性的，是一種真核菌類的微生物（fungi），在乾燥的環境下它起不到作用，但在溫度（warmth）約攝氏 32 至 35 度、濕度（moisture）及食物如糖、麵粉（food）三項有利的環境同時存在的時候，它便會活躍起來，使麵糰產生發酵膨鬆作用，讓麵糰鬆軟並帶點彈性。

一般快速乾酵母（active/instant yeast）不需要泡水，可直接混入麵粉中一起攪拌，普通的乾酵母可以依照以下的發酵過程來釀泡，同時也可知道酵母的活躍程度：

1. 乾酵母內注入暖水加少許砂糖。圖 1
2. 乾酵母開始活動，釋放出二氧化碳，冒出小氣泡　。圖 2
3. 酵母溶液已開始脹起，表示已可加入麵粉中一起攪拌。圖 3

乾酵母必須貯存於乾燥地方

工具

工欲善其事 必先利其器，以下的工具有助你得心應手炮製書中小吃。

糖環模

用於椰汁糖環（P.92）

用法：油燒熱，放入糖環模加熱數分鐘後，將熱模浸入麵漿內（不可浸過模面）。然後立刻將模放入熱油中，炸至麵漿凝固及自模鬆脫，並呈金黃色。

蘿蔔絲餅模

用於蘿蔔絲餅（P.7）

用法：油燒熱，放入蘿蔔絲餅模，當油燒至八成熱時，取出餅模，滴乾油分，加入適量蘿蔔漿，放回油中炸至離模及呈餅狀就成。

月餅模

用於雙黃蓮蓉月（P.106）

用法：新購之月餅模清洗乾淨後，先用竹籤將餅模旁之小孔通一通及用竹籤塞着小孔，注入生油浸一夜至餅模潤透為止。取出用吸油紙或油布印去表面油質便可使用。使用後，用竹籤將木模內的小麵糰清除乾淨，洗淨，要風乾後才貯藏，否則會發霉。

水晶餅模

用於水晶餅（P.54）

用法：新購之水晶餅模如月餅模般，注入生油浸一夜至餅模潤透為止，取出用吸油紙或油布印去表面油質便可使用。使用後，用竹籤將木模內的小麵糰清除乾淨，洗淨，要風乾後才收藏，否則會發霉。

雞蛋仔模

用於雞蛋仔（P.48）

用法：新購買回來之模型清洗乾淨後下點油，燒熱，抹乾淨就可使用。使用後，用清水洗淨，抹乾才收藏，否則會生銹。

圓模

用於油角仔（P.88）及椰撻（P.45）
用法：用擀麵棍將麵皮擀薄，用圓模（油角仔食譜內用圓玻璃杯口代替圓模）鈒成小圓塊，包上餡料。

撻模

用於蛋撻（P.42）、椰撻（P.45）、西米餅（P.70）
用法:餅皮放入撻模前宜掃油，這樣焗後的撻才易脫模。使用後浸熱水去油脂，並用竹籤將撻模內的小麵糰清除乾淨，洗淨抹乾才收藏，以免撻模生銹。

小砵仔

用於缽仔糕（P.56）
用法：缽仔先掃油才倒入粉漿。用砵仔蒸糕的好處是受熱均勻，避免有周邊熟、中間未熟的情況出現。

蒸籠

用於本書內的包點，如臘腸卷（P.26）、壽桃（P.109）等等
用法：將包點放入已洗淨抹乾的蒸籠內發酵，再用大火蒸。用蒸籠的好處是傳熱均勻，而且蒸籠蓋由竹編織而成，存有縫隙，故可疏導蒸氣，避免包點被倒汗水滴濕。

粉篩

用於將粉料過篩，混入空氣及將粉漿或蛋漿過篩去除粉粒和雜質。

擀麵棍

擀麵棍的用處是將麵糰擀薄。適用於葱油餅（P.10）、鹹蛋散（P.73）等。擀麵棍有不同的粗短長幼，圖中只是其中一款。

擀麵棍・餡挑

圖中的短擀麵棍多用來擀小麵糰，如生煎鍋貼（P.166）。餡挑是用來將餡料從碟子撥入皮料內，如生煎牛肉包（P.16）、上湯煎粉粿（P.13）、生煎鍋貼（P.166）等等。

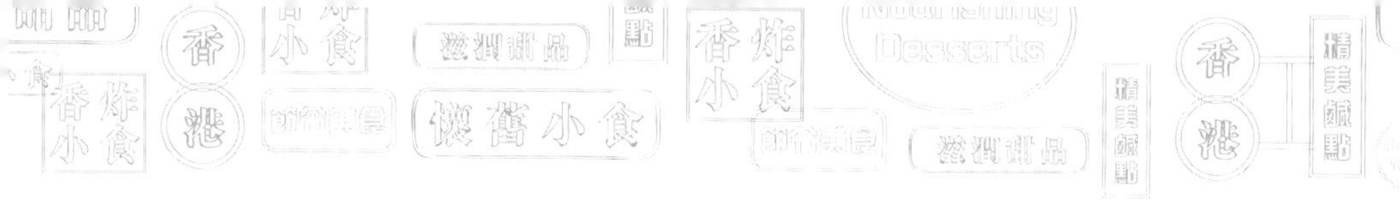

後記

《香港特色小吃》初版於 1998 年，感謝讀者們的愛戴，使之能暢銷二十載之久，這對我來説實在是極大的鼓舞。感謝萬里機構出版有限公司的邀請，讓我有機會再為《香港特色小吃》添上新裝。藉着此機緣，除了修訂了其中的十三款小吃外，我特意為大家增添了另外二十款我同樣鍾情的美食，有些是讀者們多年來要求我補上的，謝謝各方友好給我的意見。

《香港特色小吃》新訂版在製作過程中，88 款小吃，對我來説是一項體能上極限的挑戰，畢竟是已相隔了二十多年的事了，但憑着一股愛煮及愛研究的心，那擔憂都一一被我克服過了。感謝攝影師及編輯團隊，在那地獄式的製作過程裏面，我們都合作得順暢愉快，衷心感謝團隊的協作和配合。

愛煮、能煮是人生莫大的福氣，盼望再一次與讀者們分享當中的喜樂，願將最好的奉獻。書中若有不足之處，望各方不吝賜教，並繼續身心力行支持自家烹製健康、衛生及美味的香港特色小吃，同時也可以為保留本土飲食文化及承傳出一分力！

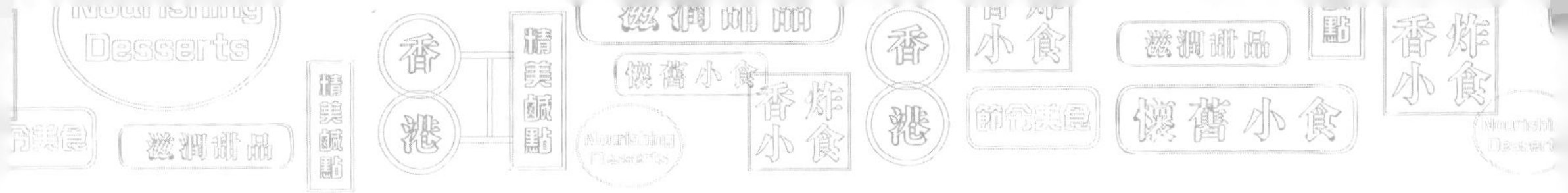

作者介紹

陳粉玉 Becky Chan

資深烹飪導師，具有四十多年的家政教學經驗，在中學教學期間，歷任家政科科主任，兼任香港教育司署（現稱教育局）轄下之成人教育中心中西式烹飪課程導師。曾任教於香港電燈有限公司之家政中心，及歷任香港教育學院（現稱香港教育大學）特約講師，經常為各類型烹飪比賽擔任評判。

她自小受家庭薰陶，對廚藝具濃厚興趣，除了教學外，還致力鑽研各式美食及出版烹飪書籍。曾為《溫暖人間》雙週刊定期撰寫素菜食譜專欄。

除了烹飪教學，對聲樂藝術亦具濃厚興趣，拜師學藝多年，曾舉辦數次個人獨唱會。致力將飲食及音樂藝術融入生活，提倡以簡單儉樸的生活提升個人精神及身心靈健康質素。

其首本專著《香港特色小吃》為香港及海外暢銷中文烹飪書籍之一，其他著作包括《快快樂樂煮幾味》、《雅淡素菜：品嘗與烹製》、《越食越聰明》、《給寶寶最好的第一口》、《學前幼兒健腦食譜》、《念念不忘潮州菜》、《心中有素》及《香港特色粥粉麵飯》。

香港特色小吃
新訂版

DISTINCTIVE SNACKS OF HONG KONG
New Edition

著者
陳粉玉

Author
Becky Chan

策劃 / 編輯
譚麗琴

Project Editor
Catherine Tam

攝影
輝

Photographer
Fai

美術統籌及設計
羅美齡

Art Direction & Design
Amelia Loh

美術設計
詩詩

Design
Venus Lo

出版者
萬里機構出版有限公司
香港北角英皇道 499 號
北角工業大厦 20 樓
電話
傳真
電郵
網址

Publisher
Wan Li Book Company Limited
20/F, North Point Industrial Building,
499 King's Road, Hong Kong
Tel 2564 7511
Fax 2565 5539
Email info@wanlibk.com
Website http//www.wanlibk.com
http//www.facebook.com/wanlibk

發行者
香港聯合書刊物流有限公司
香港荃灣德士古道 220-248 號
荃灣工業中心 16 樓
電話
傳真
電郵

Distributor
SUP Publishing Logistics (HK) Ltd.
16/F, Tsuen Wan Industrial Centre,
220-248 Texaco Road, Tsuen Wan, N.T., Hong Kong
Tel 2150 2100
Fax 2407 3062
Email info@suplogistics.com.hk

承印者
中華商務彩色印刷有限公司
香港新界大埔汀麗路 36 號

Printer
C&C Offset Printing Co.,Ltd.
36 Ting Lai Road, Tai Po, N.T., Hong Kong

出版日期
二〇一八年七月第一次印刷
二〇二四年三月第三次印刷

Publishing Date
First print in July 2018
Third print in March 2024

規格
小 16 開（240mm X 170mm）

Specifications
16K (240mm X 170mm)

ISBN 978-962-14-6770-6
Published in Hong Kong, China
Printed in China